IRAN

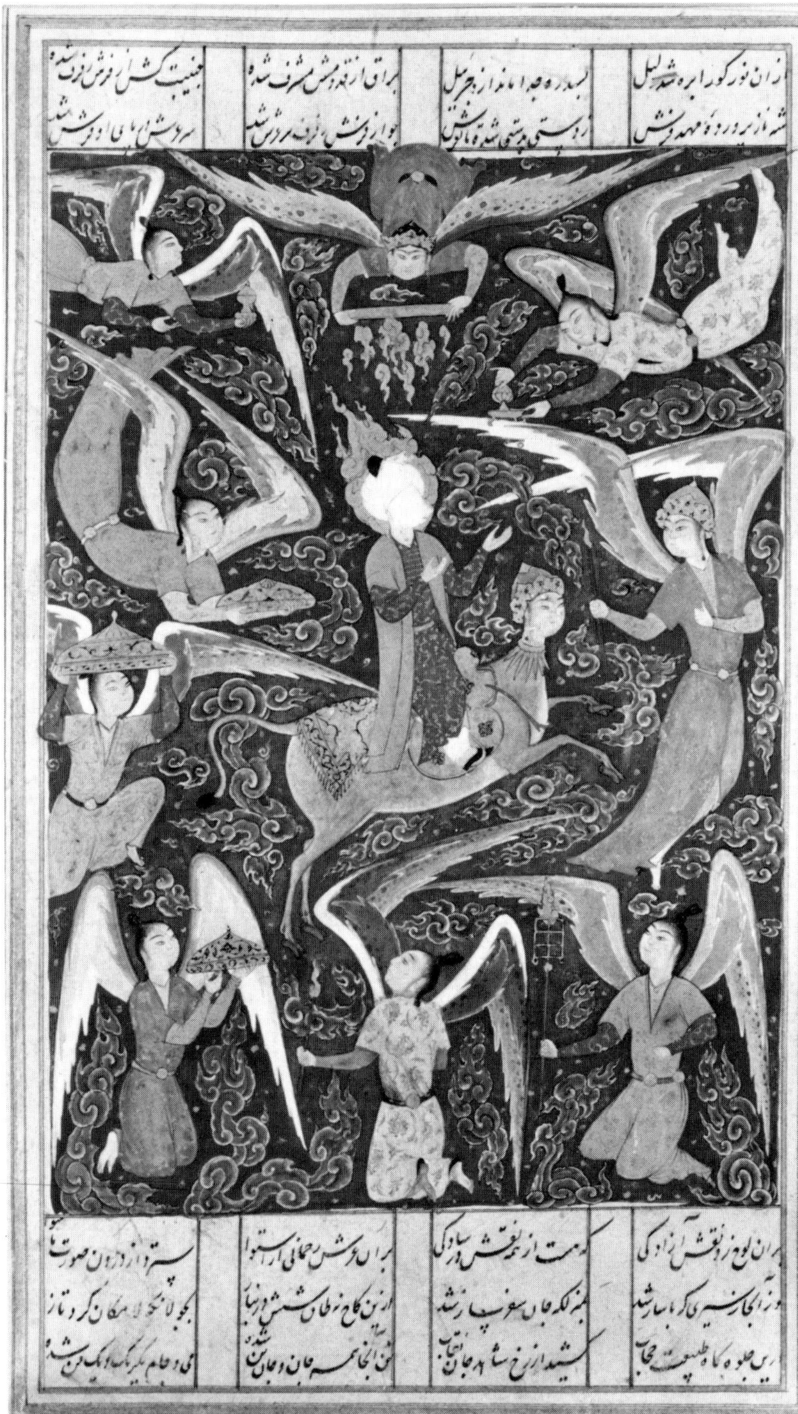

MAJOR WORLD NATIONS

IRAN

Garry Lyle

CHELSEA HOUSE PUBLISHERS
Philadelphia

Chelsea House Publishers

Contributing Author: Elizabeth Russell Connelly

Copyright © 1999 by Chelsea House Publishers,
a division of Main Line Book Co.
All rights reserved.
Printed and bound in the United States of America.

First Printing

1   3   5   7   9   8   6   4   2

Library of Congress Cataloging-in-Publication Data

Lyle, Garry.
Iran / Gar[r]y Lyle.
p.   cm. — (Major world nations)
Originally published under title: Let's visit Iran. London,
Ontario : Burke Pub. Co., 1977.
Includes index.
Summary: Describes the history, geography, and politics as well as
the social and cultural life of the people of this oil-rich
Asian country.
ISBN 0–7910–4740–7 (hc)
1. Iran.   [1. Iran.]   I. Lyle, Garry.   Let's visit Iran.
II. Title.   III. Series.
DS254.5.L93   1997
955—dc21   97–23832
CIP
AC

CONTENTS

Map   6

Facts at a Glance   9

History at a Glance   11

Chapter 1 Iran and the World   15

Chapter 2 The Land   21

Chapter 3 Ancient Empires   37

Chapter 4 Religion and Revolution   51

Color Section Scenes of Iran   57

Chapter 5 Government and Society Today   71

Chapter 6 Peoples and Ways of Life   81

Chapter 7 Resources and Economy   97

Chapter 8 A Country at a Crossroads   105

Glossary   108

Index   109

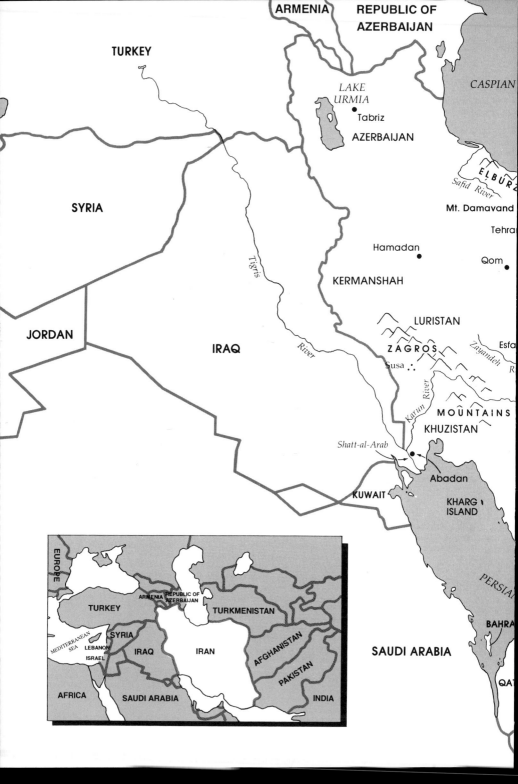

FACTS AT A GLANCE

Land and People

Official Name of Country	Islamic Republic of Iran
Area	636,128 square miles (1,648,000 square kilometers)
Population	66,100,000
Population Density	103 people per square miles (40 per square kilometer)
Capital	Tehran (population 8,700,000)
Other Cities	Mashhad (population 2,038,400), Esfahān (population 1,422,300), Shiraz (population 1,126,100)
Mountains	Zagros and Elburz ranges
Highest point	Mt. Damavand, 18,606 feet (5,671 meters)
Major Rivers	Karun, Safid, Zayandeh
Major Lake	Urmia
Official Language	Farsi (also called Persian)
Other Languages	Turkish, Kurdish, Arabic

Ethnic Groups	Persian (51 percent), Azerbaijani (24 percent), Kurdish (7 percent), Luri (2 percent), Bakhtiari (2 percent), Baluchi (2 percent), Arab (3 percent), other (9 percent)
Religions	Shiite Muslim (93 percent); Sunni Muslim (5 percent); other, including Jews, Christians, Zoroastrians, and Baha'is (2 percent)

Economy

Resources	Petroleum, natural gas, and minerals, including copper, lead, zinc, coal, sulfur, and chromite
Work Force	Agriculture (33 percent), industry and trade (21 percent), services (16 percent), unemployed (30 percent)
Agricultural Products	Wheat, barley, rice, sugar beets, cotton, raisins, dates, tea, tobacco
Industries	Oil drilling and refining, food processing, metalworking, textile and cement manufacturing
Major Imports	Machinery and transport equipment, chemicals, food, arms and ammunition
Major Exports	Petroleum and petroleum products
Major Trading Partners	Germany, Russia, Italy, Japan, Saudi Arabia, France, United Kingdom, Netherlands
Currency	Rial, divided into 100 dinars

Government

Form of Government	Islamic republic with one elected legislative body, the 270-member Majlis
Chief of State	Religious leader or religious council
Head of State	President
Voting Rights	All citizens over age 15

HISTORY AT A GLANCE

by 6000 B.C.	Settled communities practice agriculture on the Iranian plateau.
around 3000 B.C.	The kingdom of Elam flourishes in southwest Iran.
beginning 1500 B.C.	Waves of migrating peoples cross central Asia. Some settle in present-day Iran. The Medes, Persians, Parthians, and Bactrians—forerunners of modern Iranians— begin arriving around 900 B.C.
553 B.C.	Cyrus the Great, leader of the Achaemenid clan, conquers the Medes and establishes an empire that reaches into present-day Egypt, Greece, and Russia.
336–330 B.C.	Alexander the Great of Greece overthrows the Achaemenids and makes Iran (called Persia) part of his own empire. The region is fought over by Greeks and Romans for several centuries.
226 B.C.–A.D. 641	Iran is ruled by a dynasty of local rulers called the Sassanids.
7th through 9th centuries	Arabs conquer the Sassanid empire and establish the new religion of Islam throughout Iran.

by early 11th century	A Turkish dynasty called the Ghaznavids rules Iran. The Ghaznavids are replaced by the Seljuks, another Turkish people.
13th and 14th centuries	Turkish rule over Iran crumbles because of repeated onslaughts from Mongol hordes under Genghis Khan and his descendants.
1380	The Mongol leader Tamerlane establishes the Timurid dynasty in Iran. It holds together for several centuries.
16th century	A local tribe called the Safavids achieves power over much of Iran. The greatest Safavid leader is Shah ʿAbbās, who rules from 1587 to 1628.
1722	Afghan invaders conquer the Safavids. An Iranian tribesman named Nadir Shah drives them out and establishes the Afshar dynasty in Iran.
by late 18th century	The Qajars, a Turkish tribe, achieve power. They control Iran until the 1920s. During the period of their rule, European influences grow stronger in Iran.
1921–25	Reza Khan, an Iranian army officer, takes control of the government and establishes the Pahlavi dynasty, with himself as shah, or king. He takes the name Reza Shah Pahlavi.
1930s	Large-scale development of Iran's oil resources begins.
1951	The Iranian government assumes ownership of the country's oil industry, which had been principally British.
early 1960s	Under Mohammad Reza Shah Pahlavi, Reza Khan's son, Iran undergoes the White Revolution, which abolishes some traditions in favor of new land laws and Western cultural influences.
1970s	Both Communists and strict Muslim fundamentalists organize opposition to the

White Revolution. Ruhollah Khomeini, a prominent *ayatollah*, or religious leader, demands the abdication of the shah.

1979   The Islamic revolution begins. The shah abdicates his throne and leaves Iran; he dies the following year in Egypt. Under Khomeini's leadership, Iran becomes an Islamic republic, religious leaders condemn Western cultural influences, and anti-American feeling spreads through the population.

November 1979   Supporters of the Islamic revolution seize 53 people at the U.S. embassy in Tehran; most are held hostage until January 1981.

1980   The neighboring nation of Iraq invades southwestern Iran. War between the two nations continues until 1988, with heavy casualties, many of them teenage soldiers.

1989   Ayatollah Khomeini, Iran's chief of state since 1979, dies. President Ali Khamenei is appointed the new spiritual leader of the country. Ali Akbar Hashemi Rafsanjani becomes the new president and head of state. The post of prime minister is abolished.

1990   An earthquake in northern Iran kills at least 40,000, injures 100,000, and leaves 400,000 homeless.

1995   Condemning Iran as an international terrorist, the United States imposes a total ban on trade with the country.

1997   Nine Iranian women enter bids for the presidency—a first in this country where traditional interpretation of the Koran forbids women from the post. Mohammad Khatami, a relatively moderate cleric, wins the presidential election.

Ayatollah Ruhollah Khomeini returned to Iran in 1979 from exile in France to the adulation of his fervent Muslim supporters. He and his followers presided over the creation of the modern Islamic Republic of Iran.

1

Iran and the World

In southwest Asia at the crossroads of Europe, Asia, India, and Arabia lies the nation of Iran. It is about one-sixth the size of the United States, or slightly larger than the state of Alaska, and about three times the size of France. It is home to over 66 million people. Although the Iranians belong to many different ethnic and cultural groups, nearly all of them are Muslims—that is, followers of the religion of Islam.

Iran has had a long and eventful history. Villages that dotted the Iranian hills and valleys more than 8,000 years ago are among the world's oldest known settlements. Five thousand years ago, the kingdom of Elam at the head of the Persian Gulf was one of the most advanced cultures in the world. From about 3,000 years ago until the mid-20th century, Iran was often called Persia, because the inhabitants of a region called Fars, or Parsa, in southwest Iran were called Persians by their Greek enemies. Over the centuries, Persia formed part of many great empires. The armies of Alexander the Great, of Genghis Khan, and of the sultans of Turkey swept over it. These conquerors never succeeded in ruling Persia for long, but one invasion of Persia produced long-lasting results. That invasion was the onrush of Islam in the 7th century A.D. Arab warriors from the heart of present-day Saudi Arabia spread the new religion, and within a few generations Islam almost entirely replaced the native Persian religion of Zoroastrianism, which had won many followers in the ancient Roman Empire as well as in Persia.

Because the Silk Road, the most important overland trading route between Europe and China, crossed Persia, the region was a center of trade among nations throughout ancient and medieval times. Persian art and literature borrowed elements from China, India, and other sources. In turn, Persian culture was spread far and wide by the caravans of traders who carried Persian metalwork, jewelry, and woven carpets to the courts of Europe. The land that is now called Iran has thus been a world presence for thousands of years. But in the late 20th century, Iran has gained more of the world's attention than ever, for this controversial and troubled country has been torn by change and violence in recent years.

During the 1960s and 1970s, Iran and its people were propelled swiftly from their traditional way of life, based on farming and the religion of Islam, into the modern industrial world under the rule of Iran's shah, or emperor, Mohammad Reza Shah Pahlavi. The fuel that drove this economic and cultural revolution was oil. Iran has been one of the world's leaders in petroleum production since the 1950s, and the income from selling oil to other nations paid for the schools, factories, and apartment buildings ordered by the shah— and also for one of the world's largest and best-equipped armies. Then, in January 1979, a revolution of another kind occurred. A respected religious leader, or *ayatollah*, named Ruhollah Khomeini inspired a large number of Iranians who were dissatisfied with Pahlavi's rule to overthrow the shah and drive him from the country. With the support of these followers, Khomeini established a new Islamic republic in Iran, and the Iranian people were once again caught up in a whirlwind of change.

Under Khomeini's strict leadership, modern ideas and influences that had been welcomed by the shah were banned. Many Iranians who supported the shah fled the country or were killed; many others who had been exiled because of their opposition to the shah returned and took up positions of power. Iran quickly became known for its fierce hostility toward Western nations, especially the United States. In November 1979, that hostility erupted in what became known around the world as "the hostage crisis." Anti-U.S. militants in Tehran, Iran's capital city, attacked the U.S. embassy there and seized 53 U.S. citizens as hostages. The militants demanded that the shah, who was under medical care in the United States, be forced to return to Iran to stand trial. The United States responded by seizing Iranian funds that had been banked or invested in the United States and demanding the release of the hostages. Thus began a 14-month stalemate that fostered hatred and anger between the people of both nations. An attempt by the United States to rescue the captives with a secret military mission ended in disaster. The shah died in July 1980. Finally, in January 1981, the government of Algeria arranged an agreement under which the Iranians freed the hostages and the Americans released some of the Iranian funds. In the following years, however, numerous other hostages were seized and held by anti-Western terrorist and guerrilla groups supported by Iran's government. Most of these groups were based in Lebanon. Not until 1992 were the last Western hostages held in Lebanon freed.

Iran remained in the forefront of world news throughout the 1980s. A long-standing border dispute with the neighboring nation of Iraq broke out into open war in 1980, when Iraq invaded southwestern Iran. The Iran-Iraq War ground on throughout the 1980s at a cost of millions of dollars as thousands of soldiers on each side died. One target of both sides was oil shipping in the Persian Gulf, where each nation attacked ships carrying the other's oil. The

Soviet Union and the United States entered the fray in the mid-1980s when both nations allowed oil freighters belonging to the Persian Gulf nation of Kuwait to operate under their flag. As a result, both the Soviet Union and the United States sent naval forces to the gulf to protect shipping. Incidents in the gulf during the late 1980s included an accidental attack on a U.S. ship by an Iraqi bomber in 1987, the explosion of a floating mine that damaged a U.S.-registered tanker that same year, and the accidental shooting down of an Iranian passenger plane by a U.S. warship in 1988. Finally, in the summer of 1988, with their armed forces shattered and their economies deeply damaged by eight years of war, Iraq and Iran

In an attempt to extend the influence of Islamic fundamentalism beyond Iran's borders, Khomeini declared that Salman Rushdie, a British citizen and the author of The Satanic Verses, *should be murdered because strict Muslims found the novel blasphemous. These schoolgirls in Iran hold signs calling for Rushdie's execution.*

signed a cease-fire agreement developed and supported by the United Nations.

Two dramatic events in Iran grabbed headlines in the late 1980s. The first was an announcement by the Ayatollah Khomeini that shocked the world: He declared a "death sentence" against a British author named Salman Rushdie, a Muslim of Indian descent whose book *The Satanic Verses* was said by Khomeini to be an insult to Islam and to Muhammad, the prophet and founder of Islam. At this writing, Rushdie remains alive but has had to spend most of his time in hiding.

The second event shook Iranian society from top to bottom: the death of Khomeini in June 1989. Two million Iranians thronged the capital city for his funeral, a scene of chaotic, hysterical grief in which several mourners were killed and thousands were injured as the mourners trampled each other.

But fears that Khomeini's death would lead to turmoil within the political hierarchy proved unfounded. Former president Ali Khamenei was named the new, though less powerful, spiritual leader of Iran, and former parliament speaker Ali Akbar Hashemi Rafsanjani became president. In the early and mid-1990s, as the Islamic revolution appeared to lose some of its militancy, Rafsanjani focused on reviving the stagnant economy.

By the late 1990s, it was obvious that inefficient bureaucracy as well as political and ideological infighting were weakening the economy. Though anti-Western sentiments remained strong among the ruling clergy, the Iranian people showed a mind of their own in the 1997 presidential election. Rejecting a hard-liner favored by the Islamic conservatives, they chose a more moderate cleric, Mohammad Khatami, who many believed would ease some of the government's repressiveness. Like his predecessor, Khatami was known to favor better relations with the West, though skeptics thought he would have little authority over foreign policy.

The region around Golpāyegān, shown here, lies east of the Zagros Mountains at the edge of the plateau that makes up central Iran. Three mountain chains ring Iran's central plateau, much of which is desert.

2

The Land

The Islamic Republic of Iran covers an area of 636,128 square miles (1,648,000 square kilometers). It is bordered on the north by three countries of the former Soviet Union: Armenia, Azerbaijan, and Turkmenistan. Also on the northern border is the Caspian Sea, a large saltwater lake. On the east, Iran is bordered by Afghanistan and Pakistan. On the west, it is bordered by Turkey and Iraq. The southwestern border is the Persian Gulf, a long, curving arm of the sea that is fed by rivers flowing out of Iraq and Iran. About halfway along Iran's southern coast, the gulf shrinks to a narrow waterway called the Strait of Hormuz. Beyond the strait, it widens into the Gulf of Oman, which in turn widens into the Arabian Sea, part of the Indian Ocean. Iranian territory includes about a dozen small islands in the Persian Gulf.

Iran's location has shaped its history and the way its people live today. Because it lies across a major land route between Europe and Asia, for thousands of years migrating peoples, invading armies, and trading caravans have crisscrossed the area. People and cultural influences from China, India, Arabia, Russia, Turkey, and the Mediterranean met and mingled in Iran.

This 19th-century engraving depicts Tehran, the capital of Iran, nestled at the base of Mt. Damavand in northwest Iran. Tehran first became a center of commerce in the 13th century and was named the capital of Persia in 1788.

The country's location has been significant in another way as well. Just across the Persian Gulf from Iran is Saudi Arabia, where the Islamic religion was born in the 7th century A.D. When the Arabs burst out of their desert peninsula to spread their new faith, Iran was one of the first places they reached. By the 9th century, Islam had become firmly established in Iran. Today, although very few of its people are of Arab descent, Iran remains a stronghold of Islam. Most of its neighbors are Islamic nations, too: Turkey, Iraq, Pakistan, and Afghanistan. Since the collapse of the Soviet Union, Islam has also gained ground in Azerbaijan and Turkmenistan, countries on Iran's northern border. Yet Iran is unique among Islamic nations because the particular type of Islam that is its official state religion—a sect called Shia, or Shiite Islam—is practiced by less than 10 percent of the worldwide community of Muslims.

Iran's internal geography, as well as its location, has shaped its history and culture. The armies and caravans that crossed Iran did not have an easy time doing so, for much of the country is covered with steep mountain ranges and bleak deserts. Because travel within Iran was difficult, its people tended to form regional tribes and groups that remained together, spoke their own languages, dressed and lived in distinctive fashion, and cherished their separateness from all the other groups. Even today, although automobiles and airplanes have made transportation from one part of the country to another easier, Iran has many languages and many ethnically distinct population groups. One of the challenges that modern Iranian governments have faced is that of persuading the people—especially rural people in remote districts—to think of themselves first as Iranians rather than as members of the Bakhtiari or Baluchi tribe, or dwellers in the Seistan or Luristan region.

Geographic Features

Iran has three types of terrain. The center of the country is a vast, high plain, or plateau. Surrounding this plateau is a ring of mountain ranges. Coastal lowland is found outside the rim of mountains along the shore of the Caspian Sea in the north and along the southern coast.

The central plateau of Iran is one of the driest, most desolate regions on earth. It is not, however, flat and featureless. Its average height is about 4,000 feet (1,200 meters) above sea level, and the plateau is covered with low, rounded hills with wide, sloping sides. On the northwestern part of the plateau, on the inner slopes of the surrounding mountain ranges, streams and springs provide permanent water sources. In ages past, these sources were the sites of fertile oases. Now they are the sites of some of Iran's largest cities, including Tehran (the capital), Esfahān, and Qom (a religious center).

To the east and south of Tehran is one of the best-known features of the Iranian plateau: the large desert called the Dasht-e-Kavir (Great Salt Desert). Unlike other deserts around the world, the Dasht-e-Kavir consists of neither sand nor gravel. Instead, it is a crust, or in some spots a paste, of salt. The soil in this region is extremely salty, and in winter occasional rains or floodwaters from the mountains create a swampland and dissolve the salt in the soil. The waters soon dry up, leaving a glittering white crust of salt crystals. The Dasht-e-Kavir was one of the hazards of the Silk Road. The crystal surface was painful for camels and men alike to walk upon, but far worse was the ever-present threat of breaking through a weak place in the trembling crust and being sucked into the salty quicksand that lies beneath. It was said that entire caravans loaded with gold and gems perished in the Dasht-e-Kavir when they lost their way in dust storms and broke through the crust.

The southern and eastern parts of the plateau are the driest and emptiest. Southeast of the Dasht-e-Kavir lies another large desert, the Dasht-e-Lut, in the center of a vast depression. The Dasht-e-Lut is a sand desert, bare of vegetation, covered with rank after rank of peaked dunes, and torn by winds and sandstorms. The rim of mountains that surrounds the central plateau covers about half of Iran's total area. It consists of three major ranges: the Zagros Mountains in the west, the Elburz Mountains in the north, and the Khorasan Mountains in the east. In the south and southeast, smaller mountain ranges and cliffs form the edge of the plateau.

The Zagros Mountains run for about 875 miles (1,400 kilometers) from the northwestern corner of Iran south along the border with Iraq and then along the Persian Gulf all the way to the city of Bandar Abbās on the Strait of Hormuz. At their highest point, north of the city of Shiraz, the Zagros peaks reach heights of 14,000 feet (4,242 meters). For the most part, the peaks are steep and rugged, cut by mountain streams that flow through narrow gorges called *tangs*. The range is about 200 miles (320 kilometers) wide, and in its center are many deep and fertile valleys. Several ancient travel and trade routes cross the Zagros from west to east. But except for these few passes, the Zagros area is extremely difficult to travel in, and many of its settlements remain remote and isolated. To a much greater extent than the Iranians who dwell in the cities of the central plateau, the people of the Zagros have clung to old tribal ways of life.

The Elburz Mountains run in an arc from the city of Tabriz, in the northwest, along the shore of the Caspian Sea toward the northeast.

More rain falls along the Caspian coast in the north than anywhere else in Iran. The area is wet enough to support rice paddies and cypress trees, such as those shown here. The building with the cone-shaped roof is an old mosque.

Persian nobles once enjoyed tiger hunts, as portrayed in this 17th-century miniature, but increasing exploitation of Iran's natural resources and expanding settlements have made tigers all but extinct by the late 20th century.

Snow and glaciers cover the tallest Elburz peaks year-round, forming a dramatic backdrop to Tehran. Iran's highest point, Mt. Damavand in the Elburz range, is just northeast of the capital city. Its snow-capped, perfectly conical summit is 18,606 feet (5,671 meters) above sea level. Although it has not erupted since recorded history began, Mt. Damavand is a volcano, as are many peaks in the Elburz range and elsewhere in Iran. Not all of them are inactive. Mt. Taftan, a 13,000-foot (3,939-meter) volcano in southeastern Iran, periodically spews forth mud and hot gas. Volcanoes are often associated with earthquakes, and Iran has suffered its share of these destructive convulsions. In 1990, an earthquake in Ardabil, northwest of the Elburz Mountains, killed at least 40,000 people.

East of the Elburz range, the Khorasan Mountains form the eastern rim of the Iranian plateau. Not as high as the Elburz or as wide as the Zagros, the Khorasan Mountains consist of a series of rugged ranges and broad valleys. These valleys contain both grasslands for pasturing livestock and productive farmland, especially near Mashhad, which has been called the granary of Iran.

One coastal lowland area lies along the Caspian shoreline in the north. There the Elburz Mountains drop abruptly from heights of more than 10,000 feet (3,030 meters) to a narrow plain that lies about 90 feet (30 meters) below sea level. The northern slope of the mountains is quite steep, and many torrential streams and waterfalls flow down into the coastal lowland, which is marshy and moist. This lowland region is only 25 miles (40 kilometers) wide at its widest point. It was called Hyrcania by the ancient Greeks and Persians. At one time it contained several ports for shipping and trade across the Caspian Sea; today, however, although the Caspian lowland is one of Iran's most densely populated areas, trading has diminished and most people live by fishing and farming.

The remaining areas of coastal lowland are found in the south. The region called Khuzistan, in the southwest, is a flat plain lying within a triangle formed by the Zagros Mountains, the head of the Persian Gulf, and the waterway called the Shatt-al-Arab, which makes up the southern stretch of the Iran-Iraq border. This area around the Shatt-al-Arab, where the Tigris and Euphrates rivers flow together into the Persian Gulf, was the site of many early civilizations, including the Elamite kingdom, which is mentioned in Jewish and Christian scriptures.

East of Khuzistan, along the coast of the Persian Gulf, the mountains and cliffs that fringe the central plateau occasionally fall back from the shoreline. The narrow coastal lowland this creates is hot, barren, and mostly empty. East of Bandar ʿAbbās, however, the coastal ranges diminish into low hills, and a wider coastal plain called the Makran stretches from eastern Iran into Pakistan. Although the Makran is arid and hot, it has been inhabited for thousands of years by villagers whose language and culture are similar to those of Pakistan and India. These villagers fish in the Gulf of Oman and cultivate date palms along the region's few streams.

Areas within Iran are often called by regional names, some of which come from the names of tribes or peoples who live in the region—for example, Luristan, in the Zagros Mountains, is the home of the Lurs, and Baluchistan, in the southeast, just north of the Makran, is the home of the Baluchis. Some regional names are taken from local geographic features. The Khorasan region is in the Khorasan Mountains, and the region called Yazd surrounds the city called Yazd. Still other regional names, such as Hyrcania, come from ancient peoples such as the Greeks. In addition, modern Iran is divided into 25 provinces, or administrative districts, some of which use the traditional regional names. Among the most notable of Iran's traditional regions are Azerbaijan in the northwest (it has the same name as the separate nation just across the border); Kermanshah, Luristan, and Khuzistan in the west; Fars, the Makran, and Baluchistan in the south; Khorasan and Seistan in the east; and Yazd in the central plateau.

Climate and Weather

Probably the single most important fact about Iran's climate is that it is dry. Although the country has many lakes, streams, and rivers, most of them are shallow and dry up during the summer. Lake Urmia, in the northwest, is the largest permanent lake. It covers an area of about 2,000 square miles (5,200 square kilometers); unfortunately, like all of Iran's lakes, it is a saltwater lake that is useless for drinking or irrigation.

The country has three principal rivers. The largest, the Karun, flows westward from the Zagros Mountains into the Shatt-al-Arab. It is the only waterway in Iran that is navigable by boat, and then only for a short distance upstream from its mouth. On the opposite side of the Zagros range, the Zayandeh River flows southeastward from the mountains and disappears into a salty swamp called the Gavkhaneh in the Yazd region. Along its way, however, it brings life-giving water to the city of Esfahān. The third major river, the Safid, flows out of the Elburz Mountains down a steep course into the Caspian Sea. It provides water for both irrigation and hydroelectric power.

Most of Iran receives little rainfall. The deserts and the southern coast average about 5 inches (13 centimeters) a year. But although rain seldom falls along the southern coast, parts of this area, especially along the Persian Gulf, are extremely humid and uncomfortable to live in. The rest of the central plateau, including Tehran, gets a bit more rain—an average of about 12 inches (30 centimeters) a year. The Zagros Mountains average 25 inches (62.5 centimeters) yearly; the wettest region of Iran, the Caspian coastline, receives between 40 and 60 inches (100 and 150 centimeters) of rainfall each year. Summer is the driest time; at least half of all rainfall in Iran occurs during the winter months. At the highest altitudes, precipitation takes the form of snow. Water from the melted snow is a major part of Iran's water supply during the hot, dry summers.

The scarcity of water has created one of the most characteristic features of the Iranian landscape. A traveler who flies over the plateau will see rows of dark spots on the sandy surface, like dotted lines connecting outlying villages with the foothills and mountains. Each line shows the location of a *qanat*, one of Iran's underground irrigation tunnels. The dots are the holes down which workers descend on ropes to dig and maintain the qanats. These underground channels carry water from springs, streams, and melting snow in the hills into the dry central deserts for drinking and irrigation. Some of them are more than 50 miles (80 kilometers) long. Most are about 100 feet (30 meters) below the surface of the land, but the deepest are three times that depth. The qanat system is very old, and versions of it are also found in Saudi Arabia, Iraq, and Pakistan. Some working qanats in Iran have been in operation for thousands of years; in a number of villages, the job of qanat maintenance is hereditary, passed on from father to son over many generations.

In summer, Iranians divide their country into two climate zones based on altitude: *sardsir* and *garmsir*. The sardsir, or cool land, includes all areas above about 6,500 feet (1,970 meters). Here sum-

mer days are pleasant and breezy, and nights are cool. The garmsir, or warm land, is all regions at lower elevations, including the coastal lowland and most of the large cities. The traditional annual migration of many Iranian nomads takes them from the garmsir to the sardsir every summer and back down again in the winter. Some modern Iranian city dwellers follow the same pattern, moving uphill to stay with friends or relatives in the sardsir during the hot months.

Azerbaijan, in the northwest, is the country's coldest region. In winter, temperatures there can drop as low as −35° Fahrenheit (−37° centigrade). Khuzistan, in the southwest, is the hottest region, with

About 40 miles (64 kilometers) northwest of Tehran, this man-made lake provides water and electricity for the capital, and during the shah's reign it was a popular site for recreational boaters. Today, the Central Elburz National Park near this area preserves stands of ancient trees.

summer temperatures reaching 131 °F (55 °C). Tehran's temperatures range from 27° to 45°F (–3° to 7°C) in January, the coldest
month, and from 72° to 99°F (22° to 37°C) in July, the hottest month.

One other important aspect of Iran's weather is wind. The high
humidity of the Persian Gulf coast is caused by warm, damp winds
that blow across the gulf from Arabia. The low temperatures of
Azerbaijan and Khorasan in winter are caused by fiercely cold, dry
winds blowing down across central Asia from Siberia. The *shamal*,
a regular wind pattern that occurs each year from February to
October, blows northwestward along the Zagros Mountains and
the Iran-Iraq border. And a summer wind that is sometimes called
"the 120-day wind" blows westward from Pakistan, blasting the
Makran, Baluchistan, and Seistan with scorching heat and gusts of
up to 70 miles (112 kilometers) an hour.

Plant and Animal Life

Iran has more than 10,000 species of plants, some of which are found
nowhere else on earth. About 10 percent of the country is covered
in forest. In the east, the Khorasan Mountains have large stretches
of fragrant juniper forests; in the west, the slopes of the Zagros
Mountains support oak, elm, maple, walnut, pear, and pistachio
trees. In the warmer, more humid tangs, willows, poplars, and
many species of vines are found.

The most heavily forested area is the Caspian coast. Parts of this
region are almost tropical, with a thick growth of trees draped in
hanging creepers and a lush ground cover of ferns and mosses. In
a few parts of the Caspian lowland, remnants of the ancient Hyrcanian forest survive almost untouched. Scientists believe that
these stretches of woodland are among the few remaining patches
of a vast forest that covered all of Europe and western Asia many
thousands of years ago, after the glaciers of the last ice age withdrew.
The Hyrcanian forest includes rare trees such as the Caucasian wing

nut and the Caspian honey locust, as well as more common species such as the cypress, Oriental beech, hornbeam, maple, and oak. The Central Elburz National Park has been established to preserve this unusual heritage, which is the botanical equivalent of a land where saber-toothed tigers and woolly mammoths still roam.

Many once-fertile parts of Iran have been deforested or stripped of their grasslands by grazing herds of sheep and goats. Since 1962, all forests and pastures in Iran have belonged to the government, which hopes to prevent further deforestation and soil erosion by controlling land use. In the more remote mountain valleys, however, nomads and villagers continue to overgraze the pastureland and to burn trees to make charcoal for fuel.

On the plateau, vegetation is generally sparse and small, consisting of thorny shrubs, spindly tamarisk and acacia trees, dwarf palms, and hardy herbs such as giant fennel and salvia. Grapevines and palms grow in the desert oases; thickets of brush sometimes surmount sand dunes, which hold water in their depths; and lines of trees follow the courses of the qanats across otherwise barren wasteland. Many parts of the plateau, as well as the mountain valleys, are covered with a short-lived but brilliant blanket of colorful wildflowers in March or April.

Like its plant life, Iran's native animal life is both diverse and threatened by human activity. Jackals, rabbits, wildcats, and many species of small rodents are found on the central plateau. The salt desert is home to an elusive breed of wild ass. Gazelles, mountain sheep, bears, hyenas, wild boars, porcupines, and ibex, a type of large antelope, live in the wooded regions of the mountains. Panthers are still found in parts of Iran, but the Asiatic lion, once common throughout the country, is now extinct. Until recent years, tigers were occasionally spotted in the Caspian forests; it is not known whether any of them survive. Tigers have also been spotted in Baluchistan. The Asiatic cheetah and the Baluchistan bear, once

An 1899 article in Century *magazine included this drawing of eagles over the desert in Baluchistan. Eagles are still common throughout Iran.*

common, are now on the international endangered species list, along with the Persian fallow deer, the hawkbill turtle, and the Oxus cobra.

Storks, falcons, swans, and owls are native to Iran. Eagles are common, and flamingos breed and feed by the thousands in the salty wetlands along the Persian Gulf. More than 300 species of birds have been identified in the Caspian lowland; many of them are European or Russian migrants spending the winters there.

Game birds, including sand grouse, snipe, and partridge, are hunted for food throughout the country.

Biologists consider Iran unusually rich in reptile and amphibian species, with many types of frogs, lizards, turtles, snakes, and salamanders occurring in various parts of the land. The Greek tortoise is found almost everywhere. Snakes are common also, and many species are poisonous. One rare lizard, the *waran*, can reach a length of three feet (one meter). More than 200 species of fish live in the Persian and Oman gulfs, and many of them are edible; shrimps, lobsters, and turtles are also fished from these waters. Of the 30 species of fish found in the Caspian Sea, the sturgeon is the most economically important. The eggs of this large fish are known as caviar, one of the world's costliest delicacies. In recent decades, however, pollution of the Caspian by oil wells (in both Iran and the former Soviet Union) and by other industries has reduced the sturgeon population and thereby decreased the income from the export of caviar. During the 1980s, oil pollution also became a serious problem in the Persian Gulf, where attacks on wells during the Iran-Iraq War caused enormous oil slicks that threatened marine and bird life. The shah's government passed an environmental protection law in 1974, but whether or not that law is enforced in present-day Iran is unknown.

The powerful empires of ancient Persia left impressive monuments still visible in modern Iran. The tomb of Darius II (who ruled from 425 to 420 B.C.) is carved high in a cliff face and has a cross-shaped facade. It is typical of the royal tombs in Naqsh-e-Rustam, the burial place of the Achaemenid dynasty, whose members ruled Iran from the 6th to the 4th centuries B.C.

3

Ancient Empires

Archaeologists (scientists who study the remains of past cultures) have found evidence that people were living in the region that is now called Iran as long ago as 100,000 B.C. Few details are known about this extremely early prehistory, however, because the Islamic revolution in 1979 interrupted archaeological work.

More is known about Iran's history starting about 10,000 years ago. The region that now makes up Iraq and Iran was one of the first parts of the world to experience the sweeping cultural changes that marked the emergence of civilization. Farming, the domestication of livestock, pottery, metalworking, and eventually writing emerged in this area over a period of thousands of years. Although much remains to be discovered about Iran during this period, we know that by about 6000 B.C. a fairly settled culture had developed in many agricultural villages on the Iranian plateau and in Khuzistan; these were among the first such settlements in the world. Population concentrations that would one day become some of the world's earliest cities had begun to form.

By around 3000 B.C., Khuzistan was home to a culture that was more advanced than that of the plateau. Some members of this

38

This carved and inscribed block was found at Susa, the capital of the Elamite kingdom. The complex civilization of the Elamites flourished from the 4th millennium B.C. to 640 B.C., when invading Assyrians defeated the Elamites and destroyed their capital. The Achaemenids later rebuilt the city.

culture could write, and many lived in large cities, with a system of hereditary kingship and an organized priesthood. This culture became the kingdom of Elam, and it gradually spread eastward across the Zagros Mountains onto the Iranian plateau, until the Elamites built a capital at Susa, west of present-day Esfahān. Elam also included part of the region of southern Iraq that has sometimes been called Mesopotamia, along the Tigris and Euphrates rivers. The Elamite civilization lasted until the 7th century B.C., when it was overthrown by Assyrians from present-day Syria and northern Iraq. The Assyrian conquerors destroyed Susa, toppling its temples and sowing the land with salt to make it barren.

While Elam was flourishing in southwest Iran, changes were occurring in the rest of the country. Beginning around 1500 B.C., waves of migrating peoples began to enter Iran from the north. This movement was part of the centuries-long dispersal of a group of related tribes and cultures that is sometimes called Aryan and sometimes Indo-European. These peoples, whose languages shared a common origin, were the ancestors of many of the ethnic groups that today inhabit Europe, central Asia, northern India, Afghanistan, and Iran. They were fair skinned, with hair ranging from dark brown to blond in color and with brown, blue, or green eyes.

Sometime around 900 B.C., the Indo-European peoples who were the ancestors of the present-day Iranians began to arrive in Iran. Among them were the Medes, the Bactrians, and the Parthians. The Medes settled in what is now western Iran and built a capital that they called Ecbatana; today it is the city of Hamadan. Another large group settled in southern Iran, in the area that is now called Fars. At that time, it was called Parsa, or Pars, and its inhabitants were the Persians. From them came the name *Persia*, which was used for all of Iran until the 20th century.

The Achaemenids and Alexander

In Parsa, a clan called the Achaemenids rose to power and ruled the Persians. In 553 B.C., the leader of the Achaemenids conquered the Medes and united them and the Persians under a single ruler. His name was Cyrus; he is known to historians as Cyrus II, or Cyrus the Great. He and his descendants, the Achaemenid kings, created the first Persian empire.

In the 6th and 5th centuries B.C., the Achaemenids conquered the Babylonian and Assyrian empires in present-day Iraq and Syria. They then proceeded to extend their conquests eastward into modern Pakistan, Afghanistan, and Russia and westward into Turkey, Egypt, and part of Greece. Try as they might, however, the Persians could never conquer the whole Greek peninsula. Xerxes I, who ruled Persia from 486 B.C. until he was assassinated in 465 B.C., finally gave up the attempt when his fleet was destroyed by the

Greeks at the Battle of Salamis. He and his successors on the Achaemenid throne then returned their attention to their Asian empire.

The Achaemenids rebuilt the old Elamite capital at Susa and established another capital, southeast of Susa, which they called Persepolis. It was one of the grandest cities of the ancient world. The religion of the Achaemenids was Zoroastrianism, a faith that had arisen in Iran in the late 7th or early 6th century B.C. among the followers of a prophet or teacher named Zoroaster (also called Zarathustra). Zoroastrians recognized a god of good, whom they

Cyrus II, or Cyrus the Great, established the first Persian empire. He overthrew the Medean king Astyages and conquered Lydia and Babylonia. He is mentioned in the Old Testament of the Bible because he freed the Jews who were held captive in Babylonia.

Excavators at Persepolis uncovered valuable keys to the past when they discovered both clay and gold tablets covered with Persian cuneiform (wedge-shaped) writing. Unfortunately, excavation of Iran's rich historical sites came to a halt under the Islamic republic, which refuses to allow Western archaeologists access to its ruins.

called Ahura Mazda and who was represented by fire or light, and a god of evil, called Ahriman, whose symbol was darkness. The largest Zoroastrian temples were in Susa and Persepolis, and caretakers kept sacred fires perpetually burning atop the temples' tall towers.

The Achaemenid empire was the largest that had yet arisen; its size was not exceeded until the Roman Empire reached its height centuries later. But the importance of this first Persian empire was not its size. Rather, it was the opportunity it gave to population centers as far apart as Libya and India to exchange art, learning, culture, and trade goods. The horizons of civilized man took a great leap outward, from the local level to the global, under the fairly stable rule of the Achaemenids.

Persia's first period of greatness ended in the 4th century B.C., when Alexander the Great, the warrior-king of Macedonia and Greece, swept irresistibly through southwest Asia. Between 336 and 330 B.C., he defeated huge Achaemenid armies, burned the royal palace at Persepolis, took as one of his wives a princess named Roxana from a Persian territory called Sogdiana in present-day Afghanistan, and made Persia part of his empire.

Alexander hoped to make Persia a permanent addition to the Greek world by encouraging intermarriage and cultural intermingling of Greeks and Persians. Many Greeks settled in Persia, especially in the Hyrcanian region along the Caspian coast. But

This engraving depicts the forces of Alexander the Great at Granicus, site of his first victory over the Persians, in 334 B.C. Alexander completed his whirlwind conquest of Persia by 330 B.C., after defeating the armies of Darius III twice more at Issus and Gaugamela.

Alexander's death in 323 B.C. threw his hard-won empire into chaos. He left no successor, and the empire broke up as his generals fought over it. Persia—along with much of the Near East territory that had been ruled by the Achaemenids—fell into the hands of a general named Seleucus. He established a dynasty called the Seleucids, which ruled Persia for almost a century.

Parthians and Sassanids

While the Seleucids were struggling to hold on to their eastern empire, new Iranian peoples were moving into the area from the great Indo-European homelands on the steppes of central Asia. One such group was the Parni. Around the middle of the 3rd century B.C., they took over the region of Parthia, located in the center of the Iranian plateau. Although Parthia was not very fertile, being bordered by mountains and the Great Salt Desert, it included a long stretch of the Silk Road. The Parthian kings forced caravans using the trade route to pay high tolls.

Arsaces, the first of the Parnian kings of Parthia, took the throne in 247 B.C. and immediately threw off the weakened hold of the Seleucids. For the next 500 years, the Parthians continued to expand their holdings, conquering first Hyrcania, then the rest of western and southern Persia. Parthian kings such as Mithradates I and II, who ruled in the 2nd century B.C., extended the Parthian empire to the borders of Roman territory in the west and to Chinese territory in the east.

The Parthians considered themselves the heirs of the Achaemenids. They adopted Zoroastrianism—a version of Zoroastrianism, called Mithraism, even became popular in parts of the Roman Empire, where it was favored by soldiers. The Parthians were noted for their skill in horsemanship, particularly in shooting arrows from horseback. This ability gave rise to the expression "Parthian shot," meaning an unexpected blow. The Parthian empire had two capitals, one at Susa and the other at Ctesiphon, in present-day Iraq. But the Parthians were never able to grasp their empire as firmly as the Achaemenids had done. The Greeks and Romans in the west and the Chinese and Indians in the east wore down the strength of the Parthians with constant warfare, and the Parthian kingdom collapsed in the 3rd century A.D.

As the Parthians grew weaker, the Sassanid dynasty grew stronger. The Sassanids came from the region now called Fars, in southwestern Iran, the original Achaemenid homeland. By A.D. 226, they had moved outward from the kingdom of Fars through conquest and intermarriage. By the time of Shāpūr I, the first great Sassanid monarch, who ruled from 241 to 272, the Sassanid empire covered much of the territory that had belonged to the Parthian empire.

The Sassanids controlled Iran for four centuries. This period was one of internal peace, economic prosperity, and cultural enrichment. Under the Sassanids, Persian art and literature began to develop unique, distinctive characteristics. Metalwork, especially in gold, was considered the highest form of art; one gold and enamel cup from this period is preserved in a museum in Paris and is believed to be the finest surviving Sassanid artifact. The Sassanid kings also loved to commemorate their victories and their greatness in monumental rock carvings on cliff faces. Many of these huge sculptures, which show crowned kings receiving tribute, processions of mounted soldiers, or chained princes in the garb of many lands passing before their Sassanid conqueror, can still be seen throughout Iran, because crescent-shaped water moats at their base protected them from vandalism over the centuries.

Like the Parthians before them, the Sassanids waged an endless series of border wars with the Roman Empire, which after about 300 was divided into the western empire, based in Rome, and the eastern, or Byzantine, empire, based in Constantinople (now Istanbul) in present-day Turkey. But neither the Romans nor the Byzan-

An enormous Sassanid monument commemorates Shāpūr I's stunning defeat of the army of Roman emperor Valerian at Edessa in A.D. 260. Shāpūr I (shown mounted on the right) imprisoned Valerian (shown kneeling on the left) and took 70,000 Roman soldiers captive. Many of the soldiers became slave laborers and built Roman-style bridges, canals, and roads throughout the Sassanid realm.

tines directly brought about the fall of the Sassanid empire. Instead, the Sassanids were toppled by invaders from an unexpected direction—from Arabia.

The Arrival of Islam

In the early 7th century, an Arabian merchant named Muhammad from the city of Mecca experienced a series of religious revelations that he and his followers shaped into a new religion. They called this new faith Islam, which in Arabic means "submission to the will of God." In many ways, Islam was related to the Jewish and Christian faiths, for it shared their monotheism, or belief in a single god, and many of their prophets and holy figures, such as Moses and Jesus. Within Muhammad's lifetime, Islam achieved a firm basis in central Arabia. Within a few years after his death in 632, the fast-growing faith had burst the bounds of the desert peninsula where it was born, and Muslims, as its believers were called, had begun a process of expansion and conquest that would soon create an Islamic empire stretching across much of the known world, from Central Asia to Spain.

Arab warriors captured the Sassanid capital at Ctesiphon in 637 and five years later completely defeated the Sassanids. Iran was governed as a province of the Arab-Muslim empire. Zoroastrians, Jews, and Christians were permitted to follow their faiths, but they were required to pay a *jizyah*, or tax, in order to do so. Although the 8th and 9th centuries saw some rebellions against the Arab overlords and their new faith, many Iranians became Muslims to avoid the tax or because of genuine religious feeling. A number of Iranian princes and leaders adopted the Shiite form of Islam, although the most common version, called Sunni Islam, was practiced by the Arab overlords. Shiites supported Ali, a son-in-law of Muhammad, in a dispute about who was caliph, or the successor of Muhammad as leader of Islam. Shia gave rise to a number of different offshoots; Sunnism did not. One especially fanatical and secretive Shiite sect, called the Isma'iliyah, began in Egypt but established a stronghold in Iran in 1090, when its leader, Hasan-e Sabbah, gained control of Alamut Castle in the Elburz Mountains. Hasan and his successors, each of whom was called "the Old Man of the Mountain," trained religious terrorists at Alamut and similar fortresses. These individuals were called *hashishi*, meaning "users of hashish," because that narcotic drug was used in their training. The English word *assassin* comes from the word *hashishi*.

As early as the end of the 9th century, Iran was a Muslim land. It has remained one ever since. Relatively few Arabs actually settled in Iran, but the art, thought, and customs of the Arab Muslims blended with native Iranian culture. Over the succeeding centuries, most of Iran's remaining Zoroastrians left the country and settled in India. Their descendants, who still worship Ahura Mazda and his symbolic fire, continue to live there today; they are called Parsis.

Meanwhile, the Islamic conquest continued beyond the borders of Iran. By the 11th century, parts of the Christian Byzantine empire and the Arab empire had fallen into the hands of various Turkish peoples, who became Muslims. One Turkish dynasty, called the Ghaznavids, after its founder, Mahmūd of Ghazna, a warrior who achieved fame as a patron of the arts in the lands he conquered, held power in Iran throughout the early 11th century. The Ghaznavids were replaced by a ruling group called the Seljuk Turks from western Turkey. The Seljuks built a capital called Ray, just south of present-day Tehran, and ruled an empire that reached from the Mediterranean to the Gobi Desert in central Asia.

Seljuk power declined in the late 1100s, and a Turkish dynasty called the Khwarazms, based in the area east of the Caspian Sea, tried to take over Iran. But the rule of the Khwarazmshahs, as their kings were called, was brief. Iran was about to become the target of yet another invading army, one of the fiercest fighting forces the world has ever seen: the Mongol horde under Genghis Khan.

New Conquerors

In the early 13th century, Genghis's army swept south and west from the Mongolian steppes in northeastern Asia to conquer everything in its path. The horde reached Iran in 1219. The Mongol warriors destroyed Mashhad, drove the Khwarazmshahs into exile on an island in the Caspian Sea, and slaughtered the populations of entire cities. Genghis withdrew to Mongolia before his conquest of Iran was complete, but his grandson, Hulagu Khan, returned in

1256 and seized western Iran and most of Iraq. He settled in Azerbaijan, made Tabriz his capital, took the title Il-Khan, or "chief of tribal chiefs," and founded the Il-Khanid dynasty of Mongol rulers. Rivalries among powerful generals weakened the power of the dynasty in the late 1300s.

A third Mongol invasion occurred in 1380, when the war leader Tamerlane (also called Timur Lenk) conquered the Iranian plateau from the east. He founded the Timurid dynasty, which ruled Iran and Afghanistan from the city of Herat, in Afghanistan. Although the Timurid empire held together for several centuries, it eventually grew weak and disorganized. By the early 16th century, many small local dynasties had taken control of parts of Iran. The most powerful of these was a ruling family called the Safavids. They were strict

Genghis Khan led his Mongol horde west out of Central Asia between 1218 and 1224 and conquered parts of modern India, Iran, Iraq, and Russia.

This drawing of a Persian painting depicts a scene at Shah ʿAbbās's court, which was noted for its sophistication and for the production of beautiful arts and crafts.

Shiites who imposed this form of Islam on their followers. The fourth Safavid king, Shah ʿAbbās I, ruled from 1587 to 1628. Under his leadership, the Safavids drove out the last of the Timurids and achieved control over all of Iran. For the first time since the fall of the Sassanid empire in the 7th century, Iran was ruled by a dynasty of native Iranians.

Shah ʿAbbās made Esfahān his capital city and built many splendid palaces and mosques (Muslim places of worship) there. They still stand, carefully restored, and are monuments to national pride as well as architectural masterpieces. Shah ʿAbbās sponsored an expansion of Iranian arts and crafts; it was under his rule that the weaving of fine carpets, long a specialty of certain regions in Iran, became a national industry. He also firmly established Shia as the official religion of Iran. Under his rule and that of the other Safavid shahs, or kings, Iran acquired many of its modern characteristics and took pride in being free from outside oppression for the first time in many centuries.

Nadir Shah seized the throne of Persia during the period of unrest following the murder of the last Safavid shah. He led Persian armies to victory in India, where they sacked the cities of Delhi and Lahore and brought back a treasure in gems. This plunder included the Peacock Throne, a solid gold chair encrusted with diamonds, emeralds, and rubies that is now in Iran's state treasury.

4

Religion and Revolution

Like all the ruling dynasties that had come and gone before them, the Safavids eventually lost control of Iran. In 1722, invading Afghan tribesmen murdered Shah Sultan Husein, the last Safavid shah, and seized control of Iran. But this invasion sparked unrest and rebellion among those Iranians who did not intend to bow to Afghan overlords. One such Iranian was a member of the Afshar tribe, in northern Khorasan, named Nadir. He raised an army, drove out the invading Afghans, and in 1736 had himself proclaimed shah and founder of the Afshar dynasty.

Nadir Shah was one of Iran's greatest—and most notorious— rulers. His reign began triumphantly when he marched at the head of an Iranian army into the empire of the Moghuls, the Muslim rulers of northern India. He captured their capital city of Delhi and brought home an enormous treasure in precious gems and other loot (some of which today forms part of Iran's immensely valuable crown jewels). But, as his reign continued, Nadir Shah suspected everyone, including his son, of plotting to kill him and steal his treasure. He became cruel, capricious, and tyrannical. Finally, in 1747, he was assassinated by some of his own Afshar chieftains, together with members of a Turkish clan from northern Iran called the Qajars.

In the decades after Nadir Shah's death, two powerful clans vied for control of his empire: the Qajars in the north, around the Caspian Sea, and the Zands in the south, along the coast. A small remnant of the Afshar state remained intact in Khorasan. Then, in 1794, Āghā Mohammad Khān of the Qajars defeated his Zand opponent in the south. Two years later, he defeated the remaining Afshars in Mashhad, where he tortured Shah Rokh, Nadir Shah's blind grandson, to death in an attempt to unearth hidden treasure. He had himself proclaimed shah of all Iran, beginning more than a century of Qajar rule.

The most significant feature of the Qajar era in Iran was the increased European influence. Qajar rulers allowed Russia, Eng-

Reza Khan was an able, tough lieutenant colonel in the Cossack Brigade of the Persian army when he seized power in 1921 with the aid of the British, who provided arms and support in return for his promise to further their oil-drilling interests. By 1925, he had made himself shah and accelerated the pace of Westernization in his nation.

land, and other countries to open embassies, mines, and trading companies in Iran in exchange for cash payments, called subsidies, to the royal treasury. The shahs spent this money lavishly on themselves and their wives; they also traveled in Europe and began to copy Western and Christian ways, which made them increasingly unpopular with the Iranian people, most of whom were uneducated tribespeople and strict Shiites. After the beginning of the 20th century, the shah was forced by popular pressure to give the country a constitution and an elected legislature, called the Majlis. World War I (1914–18) brought famine, national bankruptcy, and increased pressure from England and Russia, both of which wanted to profit from Iran's mineral resources and its strategic position between Europe and Asia. Despite these problems, the years just after the war saw the rise to power of the man who created modern Iran.

The Pahlavis

In 1921, a forceful and intelligent Iranian army officer named Reza Khan stepped into the chaos of Iran's government and took control of the armed forces. He controlled Iran for the next two decades. He forced the government to name him minister of war, then prime minister. In 1925, he deposed the last Qajar shah and had himself crowned Reza Shah Pahlavi.

Under Reza Shah's rule, the Westernization that had begun under the Qajars continued, but it affected all of Iranian society, not merely the privileged classes. The influence of religious leaders over the state grew weaker. The educational and legal systems were rebuilt on European models. Women were given the right to vote for members of the Majlis and were freed from the Islamic obligation of wearing the head-to-toe *chador*, a combination robe and veil, at all times. The economy was also modernized when oil deposits were located and a British company began drilling wells and building refineries in Iran. The country's first railroad opened for service in 1938. In 1935, the shah asked world governments to call his

Women were given the right to vote and freed from the obligation to wear the chador under Reza Shah in the 1930s; in 1979 these women chose to wear the chador and voted for the establishment of an Islamic republic, which restricted their freedom.

country Iran rather than Persia. The new name, which had long been used within the country as an alternative to Persia, is a version of "Ariana," which means "country of the Aryan people."

World War II (1939–45) brought more changes. The Allied forces, especially Britain and the Soviet Union, wanted to ensure that supplies of Iranian oil would continue to reach the front, so both nations sent troops into Iran to prevent Nazi Germany, their enemy, from gaining control there. Reza Shah, however, favored Germany, partly because he resented the British and Soviet intrusions into his country and partly because many Germans were living and working in Iran at the time. Pressure from the Allies forced Reza Shah to abdicate—that is, to give up his throne—in 1941. He left Iran, and his son, Mohammad Reza Shah Pahlavi, ascended the throne.

After the war, the occupying British forces left Iran, but the Soviet forces remained and attempted to take control of the Azerbaijan region. Protests by the shah, the British government, and the United

Nations finally caused the Soviets to withdraw beyond the border in late 1946.

The postwar years were a time of surging economic growth and social change. In 1951, the Majlis voted to nationalize—that is, to make the state the owner of—Iran's oil industry, which was dominated by a British-owned company. Mohammad Mosaddeq became prime minister and succeeded in nationalizing the oil com-

During the 1950s, the shah enacted many programs designed to reform Iranian society, including several measures that broke up large estates and divided the land among peasants. Here, a peasant who has received a grant of land demonstrates his gratitude to the shah. Not all Iranians were so thankful—many felt the shah's reforms showed disregard for their traditional society and religion.

pany. Since that time, the Iranian government has retained close control of its country's oil resources.

Unfortunately, however, the output of the oil wells and refineries declined under Iranian management, and the nation's economy, which floated on petroleum, began to sink. Mosaddeq's popularity also fell. In August 1953, the shah tried to have him removed from office but failed—instead, the shah was forced to flee Iran in fear for his life. Four days later, however, the shah returned with the forceful backing of the United States and other Western powers, which wanted to be sure that Iran's oil production would not suffer. The United States gave Iran $45 million in aid, and the shah's government denounced Mosaddeq as a traitor.

During the 1950s and 1960s, the shah grew more powerful and the Majlis less powerful. Oil income increased steadily and paid for new schools, hospitals, public buildings, roads, and other developments. In the early 1960s, the shah introduced what he called the White Revolution in order to change many aspects of traditional Iranian life. Women were given the right to hold public office, large estates that had belonged to wealthy or powerful families for generations were broken up into small farms and distributed to peasants, and education was made a national priority. In many ways, the shah's reforms were designed to benefit the Iranian people; many Iranians, however, felt that their traditions and religious beliefs were being disregarded. Furthermore, the shah's methods became increasingly tyrannical. Freedom of speech was limited. Communist political and labor organizations were banned. In 1975, Iran became a one-party state when the shah outlawed all political parties other than his own National Resurgence party. Religious and political leaders who criticized the shah were jailed; some, including the Ayatollah Ruhollah Khomeini, left the country. The shah's secret police force (trained by Israeli agents and the U.S. Central Intelligence Agency), called SAVAK, was hated and feared as a symbol of cruelty and repression and was

(continued on page 65)

SCENES OF
IRAN

58

Overleaf: Iranian women cloaked in chadors, as centuries-old Muslim tradition demands, pass by an enormous modern oil refinery. Such contrasts are common in the Islamic Republic of Iran.

The city of Esfahān boasts some of the loveliest mosques in Iran. The dome of the Sheikh Lutfullah Mosque is reflected in the calm, still pool before it.

A close-up of a mosque dome in Esfahān reveals its intricate decoration. Swirling figures derived from animal and floral motifs surmount a band of Arabic inscriptions above complex geometric patterns. Two minarets, similarly covered with brilliant tile, appear to the left of the dome.

Women pray inside the Friday Mosque in Esfahān. Elaborate geometric tile patterns characterize the interior as well as exterior decoration of mosques. As is Muslim custom, women pray in a different area than male worshipers do. In this case, a sheet provides a separating screen.

Young weavers in Tabriz work to complete one of the carpets for which their city is renowned.

Iranian carpets such as this one fetch tens of thousands of dollars from buyers around the world. Many such carpets become valued heirlooms and are resold at high prices as well.

Craftsmanship and fine artistry have a long tradition in Iran. This detail from a fresco painted in the time of Shah ʿAbbās I reveals the attention to minute detail found in Persian painting. Frescoes were often composed of dozens of such figures.

An oil rig near the Iran-Iraq border in Khuzistan spews fire and smoke into the sky. During the Iran-Iraq War, oil fields were a major target of bombings, as each side sought to cripple the other's economy.

A young soldier crouches beside an Iranian flag and watches for enemy Iraqi forces. Thousands of teenage soldiers died in the war.

Concrete apartments built during the shah's rule crowd modern Tehran. Money from oil sales fueled a construction boom in the 1960s and 1970s.

This Iranian youngster can look forward to a turbulent future. Iranians must struggle with the demands of remaining a theocratic, Islamic republic in a world where few such states exist.

(continued from page 56)

responsible for mysterious disappearances and torture. The shah's government spent an enormous amount of money on arms, military equipment, and costly state ceremonies.

Opposition to the shah grew during the 1970s, especially among two groups: left-wing or Communist-inspired students and intellectuals on the one hand, and Muslim fundamentalists, or believers in the strictest possible interpretation of Islamic doctrine, on the other. One of the most prominent spokesmen of the opposition movement was Khomeini, a religious teacher and leader, who was living in Paris. Khomeini especially hated the shah's close connection with the United States, a nation he regarded as impure and evil. Khomeini's speeches and books were smuggled into Iran by his followers. There they helped to fuel the flames of resentment against the shah.

The Islamic Revolution

In 1978, demonstrations against the government occurred in Tehran and other cities, and the shah imposed martial law that September. But the antigovernment movement grew stronger the more firmly the shah tried to repress it, and the rioting could not be stopped. In January 1979, the shah fled the country, never to return. Two weeks later, Khomeini flew in from Paris and was greeted by rapturously chanting crowds. From that moment he was the real leader of Iran, although several presidents held office under his control. His white beard, black turban, bushy eyebrows, and piercing eyes swiftly became known around the world: To many Muslims, he was a symbol of an Islamic renaissance; to many Westerners, he was a dangerous, intolerant madman.

Khomeini's followers formed the Islamic Revolutionary party (IRP) and proclaimed Iran an Islamic republic. The country became a true theocracy—that is, a state in which the official religion is also the supreme governmental authority. Reforms, fashions, Western trends, and anything else not in harmony with strict Shiite tradition immediately became exceedingly unpopular or even illegal. The chador became mandatory for women, *ulemas* (councils composed of local *mullahs*, or respected religious leaders) took over the courts, and modern divorce and property laws were canceled. In the midst of this Islamization, many Iranians who had been educated in the West, whose families or businesses had been associated with the shah, or who simply had come to enjoy Western freedoms fled for Europe and the United States. The country lost many doctors, teachers, and technicians.

The 10-year reign of Khomeini was marked by many dramatic events: the U.S. hostage crisis in 1979–81; the execution of Foreign Minister Sadegh Ghotbzadeh in 1982 on charges that he had planned to assassinate Khomeini; thousands of deaths and arrests due to fighting among different factions within the IRP and to urban riots; the persecution of religious minorities, such as Jews and Christians, most of whom fled the country during the 1980s; and growing discontent among many Iranians with the tyrannical, repressive, sometimes violent rule of the ayatollah. In foreign affairs, the Khomeini regime endured some disappointments. Although its staunchly anti-Western position never weakened, Iran did not achieve the position of leadership in the Islamic community for which Khomeini had hoped. Iran remained on hostile terms with some Arab Muslim nations, such as Egypt and Saudi Arabia, because Khomeini felt that these countries were too eager to get along with the West and with Israel. Iran's closest allies were Libya and Syria, generally regarded by Westerners as the most extreme, unreliable, and warlike of the Muslim nations.

But Iran's biggest problem was its war with Iraq. Disputes and border clashes between the two nations had flared up periodically since the early 1970s. In 1980, Iraq demanded full control over the Shatt-al-Arab and invaded Khuzistan. Iraqi forces seized the Iranian town of Khorramshahr and the large oil refinery at Abadan, both on the Shatt-al-Arab. Khomeini immediately declared all-out war on Iraq and called for army enlistments. The war machines the shah had built were put to use against Iraq, and by 1982 the Iranians had driven the Iraqis out of Khuzistan and had launched their own attack on the Iraqi city of Basra. As the war dragged on year after year, however, neither side could gain a clear victory. The death toll mounted into the tens of thousands on each side, and both Iran and Iraq, each of which maintained an army of more than 600,000, began having trouble keeping their forces fully staffed. One of the saddest aspects of the war was the enlistment of boys as young as 11 or 12 years old who died by the thousands in what each side claimed was a "holy war." The United Nations tried for years to arrange a cease-fire, or an end to hostilities.

Ayatollah Ruhollah Khomeini returned to his rapturous followers in Iran in January 1979. During the ayatollah's 15-year exile in Iraq and France, he urged his supporters to revolt against the shah through daily tape-recorded speeches smuggled into Iran on cassettes.

Two Iranian soldiers check the barbed wire enclosing 2,000 captured Iraqi soldiers in Ahwaz in southwest Iran. The prolonged, costly Iran-Iraq War claimed tens of thousands of lives on both sides.

Finally, in the summer of 1988, both nations agreed to the UN proposal. Khomeini, who had once sworn not to stop fighting until Iraq was ground into the dust, said on Tehran radio, "Making this decision was more deadly than drinking poison. I submitted myself to God's will and drank this drink for his satisfaction." The cost of the war in lives and money cannot be fully known, for both Iran and Iraq have maintained great secrecy on the matter, but UN experts estimate that a total of at least half a million people died in the fighting. Paying for the war drove both countries deeply into foreign debt, and the relentless bombings of each other's oil wells, refineries, and tankers and the loss of farmers and other workers crippled both the Iranian and Iraqi economies.

Khomeini's death in 1989, when he was in his late eighties, created a power vacuum in Iran. In the months immediately after the death of the ayatollah, Ali Akbar Hashemi Rafsanjani, who became Iran's president, emerged as the new political leader. Rafsanjani diminished the influence of fundamentalist and revolutionary factions and placed greater emphasis on economic development. In the early years after the Islamic revolution, key industries such as oil, utilities, and transportation had been nationalized—that is, taken over by the government. Rafsanjani attempted to liberalize the economy and reduce government control. Continued economic problems, however, somewhat revived the power of hard-line factions associated with Ali Khamenei, the spiritual leader who had replaced Khomeini.

Rafsanjani, who served the constitutional limit of two terms, was succeeded in 1997 by Mohammad Khatami, a relatively moderate cleric and former culture minister. On the domestic front, Khatami was expected to expand industry privatization (transfer of control back to private companies) and encourage greater social and political freedom. But it was not clear whether he would have enough power to influence international relations to a significant degree. By the late 1990s, the tug-of-war between conservative ideologists and moderate pragmatists had become a fact of political life in Iran.

The man in the rear, a soldier wearing a traditional turban and carrying a modern machine gun, symbolizes the combination of religious, political, and military authority that marks the government of modern Iran. In front of him, the boy in fatigues may be contemplating being sent to fight for his country, which has enlisted soldiers as young as 11 years old.

5

Government and Society Today

Iran is governed under a constitution that was written by Khomeini and his supporters and accepted by the people's vote in 1979. This constitution defines Iran as an Islamic republic in which the official head of state is the president, who is elected by popular vote every four years. All citizens over age 15 may vote.

When Khomeini was alive, the president of the republic was largely a figurehead. The constitution called him the head of state, yet he had little real power. The prime minister, originally the head of government, had even less power than the president, and the position of prime minister was abolished in 1989. During President Rafsanjani's two terms of office (1989–97), the situation changed somewhat. Rafsanjani managed to strengthen the presidency.

Today, with the legislature's approval, the president appoints the members of the Council of Ministers, a group of advisers whose function is similar to that of the cabinet in the U.S. government— that is, they oversee the day-to-day administration of various branches of government administration, such as labor, economic planning, commerce, social welfare, and the like.

The constitution also created a position that was filled by Khomeini during the 1980s: that of *valiy-e faqih*, or supreme spiritual leader, who is also the nation's chief of state. The spiritual leader is supposed to be elected by religious experts or scholars chosen by the people. (In the event that a single valiy-e faqih cannot be chosen, the constitution permits the election of a council of three to five highly regarded Muslim theologians to fill the position.) The valiy-e faqih is the ultimate authority on government as well as on religion. He commands the armed forces and approves all candidates for president and the Council of Ministers. He is aided by a Council of Guardians, which consists of six religious leaders whom he appoints and six lawyers appointed by the legislature. These guardians scrutinize all laws, court decisions, and government actions to make sure that they meet the principles of Shiite Islam and of Iran's constitution. When Khomeini was valiy-e faqih, his authority was complete. After his death, however, no single religious leader possessed stature and popular support comparable to his; there was no one to assume his position of supreme leadership. Ali Khamenei became the new valiy-e faqih, but it was understood by both Iranians and outsiders that he would be far less powerful in that position than Khomeini, who had been widely respected as a Muslim scholar and teacher for many years before the Islamic revolution began.

Iran's legislature, sometimes called its parliament, consists of a single body of 270 representatives who are elected by the people for 4-year terms. This body, called the Majlis, or National Consultative Assembly, makes the country's laws under the close supervision of the Council of Guardians.

Local government is administered by the governors-general of the 25 *ostans*, or provinces. These governors, as well as other district officials, are appointed by the central government. The day-to-day job of local administration is shared between government officials and religious bodies called *anjumans*, or provincial councils, which are made up to a large extent of mullahs. Through these councils, the revolutionary government is able to keep a close watch on the activities of all citizens, and the mullahs are free to enter any home or business at any time to check for violations of Islamic law, such as the drinking of alcohol or the appearance of women without veils. Such violations are strictly punished.

Under the constitution, Iran's justice system is based on the Shari 'ah or body of traditional Islamic law. The judiciary system consists of a Supreme Court, which meets in Tehran, and a series of lower courts across the country. The highest-ranking judicial body is the High Council of the Judiciary, which is made up of four

Rafsanjani (center) consolidated his power as president of Iran after Khomeini's death in 1989. However, he had been active and influential in the government for more than a decade. This photograph shows him holding a news conference regarding the Americans taken hostage in the U.S. embassy in 1979.

Under the shah, U.S. assistance enabled the government to establish 375 tent schools to educate Iran's population of nomads. These unique schools moved along with the tribes, giving them access to daily instruction.

mujtahids, as experts in Islamic law are called. They establish legal policies and oversee the operation of the courts. The penal code was revised in 1983 to include traditional punishments called *qisas,* or "retributions," which had been banned under the Pahlavi dynasty. In the qisas system, prisoners convicted of certain crimes are punished by having an eye put out or a limb cut off, or sometimes by being beheaded or executed in some other manner. The punishment may be carried out by the state or, in cases of punishment for murder or other violent crimes, by the family of the injured party.

All Iranian men must serve in the armed forces for two years. These forces consist of an army, an army reserve, a navy, and an air force. In addition, a special force called the Revolutionary Guard Corps is responsible for maintaining order within the country and quelling any disagreement with the government of the revolutionary leaders. By the mid-1990s these forces had a total of over half a million men.

As far as political operations are concerned, the IRP was made Iran's official party after the revolution. Nearly all members of the Majlis, the justice system, and the government belonged to the IRP. But that party was dissolved in 1987, leaving Iran with no effective political parties. Today, several dozen parties are licensed to exist, but are not yet openly active. A number of armed political groups, such as the People's Fedayeen, have been rigorously suppressed or outlawed.

Education, Health, and Social Welfare

Like many aspects of life in Iran since the Islamic revolution, the status of the country's educational system today is something of a mystery. The country's tradition of education at Islamic schools and universities for men goes back to the early centuries of Islam, when Persia gained renown as a center of learning. Such learning, however, was restricted to the wealthy or to those who planned to devote their lives to religious study. Education in the modern sense came late to Iran and was introduced by the shah.

In 1943, education through the fifth grade was made compulsory (that is, it was required by law) for all Iranians. It was at this time that schooling for girls and women—a notion quite foreign to both Iranian tradition and Islamic culture in general—was introduced. After 20 years, however, the literacy rate (the number of adults who could read and write) remained low. In 1963, the shah's government tried to improve the nation's literacy by sending squads of trained young people into the countryside to teach reading and writing. This effort was slowed by the revolutionary upheaval, but by 1994 it was estimated that about 72 percent of the adult population was literate.

At the time of the 1979 revolution, Iran had 16 universities, the largest of which was the University of Tehran, founded in 1934. All of these schools were closed by the revolutionary council. They were gradually permitted to reopen under the close supervision of mullahs from the IRP, except for the departments of law, philosophy, and certain other disciplines that were declared to be foreign to the Islamic system. Today each university is run by a regent or board of regents appointed by the state. The major universities are at Tehran, Tabriz, Mashhad, Esfahan, and Shiraz. The country also has about 50 smaller colleges, including institutes for training teachers, and about 40 technical schools.

In the early 1980s, Iran had about 45,000 primary schools for children from ages 7 through 11 and about 9,300 secondary schools

This young woman spoke in Washington, D.C., in 1985 about human-rights abuses under the Khomeini regime. She lost a leg and endured horrible conditions when imprisoned at the age of 20.

for children from ages 12 through 18. Little is known about the number of students or the quality of education in these schools today, but it is certain that many teachers left Iran after the revolution began and that many others have been imprisoned or dismissed for counterrevolutionary activities. All schools are operated by the government's Ministry of Education and Training. Khomeini did not do away with education for women, as some had feared he might, but he did impose sexual segregation on Iran's schools, so that girls and boys are now educated separately at all levels. All education is free, but students who wish to attend a university must agree to work for the government later for a number of years equal to the time spent at the university.

One area in which the shah brought dramatic improvements to Iran is that of health care. Smallpox, cholera, malaria, tuberculosis, venereal disease, typhoid fever, amoebic dysentery, and other deadly conditions that had plagued the country for centuries were brought under control. Hospitals were built in many cities during the 1960s, and a health corps of young physicians and trained secondary-school volunteers was formed in 1964 to take basic medical care into remote villages and mountain valleys. In 1979, Iran had 1 doctor for every 2,282 people and 1 hospital bed for every 660. But many doctors and nurses, most of whom had been trained in the West and were not sympathetic to the Islamic revolution, fled Iran when Khomeini came to power. Health conditions in the country reportedly deteriorated during the 1980s, especially in rural areas, where the shortage of doctors and medicine was severe. Recently, however, the nation seems to be making progress again. Estimates say that Iran now has 1 doctor for each 1,600 people and 1 hospital bed for every 650. Average life expectancy is about 69 years for women and 66 for men.

The constitution of 1979 established some programs to provide social-welfare benefits, including insurance for those who are ill or otherwise unable to work. Iran also has an old-age pension and

Prominent members of the Supreme Judicial Council attended a memorial service for Khomeini in June 1989. From left to right: Ali Khamenei; Khomeini's brother, the Ayatollah Pasandidi; Khomeini's son, Ahmet Khomeini; and Rafsanjani, who was then the speaker of parliament.

organizations to provide for the well-being of the families of war casualties.

The violation of individual civil rights under the shah's rule and by SAVAK was one source of the outrage that sparked the Islamic revolution. Ironically, however, civil rights violations continued under Khomeini's rule; some sources say that they were worse than under the shah. All criticism of the IRP was ruthlessly suppressed, and the government conducted arrests, trials, and executions without regard for the processes required by law. As of 1982, according to Amnesty International, the worldwide civil rights organization, Khomeini's government had been responsible for the death of at least 4,658 political prisoners. The IRP was also accused of torture and of persecuting members of religious minorities, such as Jews. Soon after the Amnesty International report was published, Khomeini freed 8,300 prisoners, but many political prisoners remain in jail.

Civil rights abuses, including "disappearances" and blatant murder, also persist beyond Iran's borders. In 1997, a German court implicated Iran's top leadership in the 1992 slaying of four dissidents in Berlin—the sort of act traditionally blamed on revolutionary terrorist groups. This historic ruling ruptured relations between Iran and its biggest trading partner, and it fanned anti-Western sentiment throughout Iran.

Vendors and shoppers crowd the streets of Tehran outside the Shah Mosque, built by Mohammad Reza Shah Pahlavi. Iranian cities have grown tremendously in the second half of the 20th century as villagers have flocked to them in search of jobs.

6

Peoples and Ways of Life

About 51 percent of Iran's people belong to the ethnic group called either Persian or Iranian. They are considered to be direct descendants of the Aryan tribespeople who migrated into Iran from the steppes of central Asia some 3,000 years ago. The majority of these ethnic Persians live in Tehran, Esfahān, Shiraz, and the other large cities of the northern and western plateau. They speak Iran's official language, Farsi, which is sometimes called Persian. This language comes from the original Indo-European language of the Aryan peoples, although many words from Arabic and other languages have been added to it over the centuries. Farsi is written with a version of the Arabic alphabet.

In addition to its Persian population, Iran has many other, smaller ethnic groups, some of whose members speak their own languages. The largest of these groups is the Azerbaijani, who make up 24 percent of the population. They live in Azerbaijan, in the northwest, and they are descended from Turkish peoples who settled in the area after the 10th century. They speak Azari, a dialect of the Turkish language. The Azerbaijani are farmers, herders, and traders. They have lived in their settled towns and villages for centuries and tend to keep to themselves, although they are a stable and prosperous element within Iranian society.

The Zagros Mountains are the home of several tribal peoples who together make up a significant percentage of Iran's population. Chief among them are the Kurds (seven percent), the Luri, or Lurs (two percent), and the Bakhtiari (two percent). These peoples speak languages of their own that are derived from the ancient Indo-European language and are similar to Farsi, although they are distinct from it.

The Kurds, traditionally a nomadic herding people, live in the northern part of the mountain range; their ancestral territory also extends into parts of Turkey, Syria, and Iraq. For many years, the Kurds have resisted becoming part of these modern nation-states. They maintain that they should be free to cross the borders with their herds of goats and flocks of sheep as they have done for centuries. After World War I, Kurdish nationalism—the idea that the Kurds should have a state of their own—gained some popularity, but it was crushed by the governments of Iran and of the surrounding countries. The Kurds opposed the government of the shah, who relocated many of the most troublesome of them to the eastern region of Khorasan, and they continue to resist government control. In 1991, about a million Kurds fled to Iran to escape Iraqi forces after the Persian Gulf War. Today, the border between the two countries is officially closed to these wanderers.

The Luri live in the central part of the range, in the region known as Luristan. The Bakhtiari, to whom they are closely related, live in the mountains west of Shiraz. Both speak the Luri language. Like the Kurds, these are peoples of nomadic traditions. They live in round black tents made of goat hair and pasture their flocks on the high mountain slopes during the summer and in the milder valleys during the winter. Beginning in the 1960s, however, efforts were made to end their nomadic way of life, which the shah considered primitive and backward. Today the government continues a program, begun by the shah, to resettle the nomadic tribes in farming villages. It is not known how many families continue to follow the old ways of the Kurds, Luri, and Bakhtiari.

One feature of the traditional life of the nomadic peoples of the Zagros deserves special attention: Women have been given more freedoms and higher status among these tribespeople than among the other Persian peoples. They were not required to wear veils or

Kurds display their traditional tribal costumes. Fiercely independent, the Kurdish people have resisted government control for decades.

In the culture of several nomadic tribes, women were granted higher status than they were in traditional Persian culture. This young girl of the Shahsevan tribe displays her elaborate finery and holds an open book, perhaps indicating that she is able to read—a highly unusual accomplishment for women and men of her tribe before the modernization of education that the shah introduced.

chadors and were permitted to mingle freely with men—even, in some cases, to govern tribes and lead war parties. But throughout the rest of Iranian society today the position of women is dictated by Islamic traditional law, which gives them a status lower than men and requires them to be subject to their fathers, husbands, or brothers in financial and domestic matters. Although many women in Iran work and women are permitted to vote, they are still expected to remain obedient to men and silent in their presence.

Another two percent of Iran's population consists of Baluchis, a tribal people of the Baluchistan region in the southeast. They live in huts made of willow branches, and although they were once nomads, today they tend to be farmers rather than herders. Their language is of Indo-European origin.

Iran's Arabs make up about three percent of the population. Most of them live on the Persian Gulf islands or in Khuzistan, which has been called Arabistan at various times in Iran's history. Today many of these Arabs work in Iran's oil industry, but some still keep herds of cattle, sheep, and camels and tend groves of date palms. They have close cultural and familial ties with peoples in Iraq and Saudi Arabia; some of the Arabs have even demanded that Khuzistan become part of Iraq. This demand was among the causes of the Iran-Iraq War.

Many other ethnic groups live in Iran. Among the most important of them are the Qashqai, a nomadic people of the southwestern highlands who speak a Turkish dialect and have often rebelled against both the shah and the revolutionary government. The Turkomans of Khorasan, another Turkish-speaking nomadic group, are related to peoples of neighboring Turkmenistan. The Khamsheh, Mamasani, and Shahsevan are other migratory groups. The Armenians are descended from the ancient country of Armenia, part of which became independent again after the breakup of the Soviet Union. The Armenians in Iran are concentrated in Tehran, Esfahān, and Azerbaijan. They have attempted to preserve the Armenian language and customs in their communities; most of them are businesspeople. In the southeast, in the Makran and southern Baluchistan, lives a scattering of people called Brahui, whose language belongs to the Dravidian language family (the source of the languages of southern India) and who are culturally and ethnically related to the Indian people. A few Syrians, Afghans, and Pakistanis also live in Iran.

Until the 1979 revolution, some foreign-educated Iranians and international businesspeople spoke European languages—mainly English, French, German, and Russian—in Iran. Today the use of these languages is discouraged. Arabic is more commonly spoken than it used to be and is granted respect because it is the language of the Koran, the sacred book of Islam.

Despite its diversity of peoples and cultures, Iran has almost no religious diversity—98 percent of its people are Muslims. And nearly all of them are Shiite Muslims, members of a form of Islam that is greatly outnumbered around the world by the Sunni Muslims, who make up only five percent of Iran's population. (The two schools of Islam split centuries ago in a dispute over the succession of religious and political leadership among the followers of Muhammad.) Iran is the only country in the world in which Shia, or Shiite Islam, is the state religion. But although the Shiites and the Sunnis have been enemies at times over the years, both groups are far more closely allied with all their fellow Muslims than with any non-Muslims.

A sea of believers bows toward Mecca at the Friday prayer meeting at Tehran University. Iran was overwhelmingly Muslim before the establishment of the Islamic republic, and religious diversity has been further reduced since 1979.

Members of a few other Islamic sects live in Iran. Sufism, a school of Islamic mysticism, had many followers in medieval Persia, and a few Iranians still practice its rituals, some of which involve beating or stabbing oneself or achieving a trance through prolonged dancing. Ismailism, the sect of the medieval assassins, lingers on in Iran and other Islamic countries; the Druze, a militant Islamic group that has been involved in many acts of terrorism, is part of this sect.

Khomeini's regime was not noted for its tolerance of any non-Islamic religion, although the 1979 constitution grants religious freedom to Christians, Jews, and Zoroastrians. The Christian population numbered about 333,000 in 1980, mostly Armenians and Azerbaijanis. That same year, there were about 30,000 Zoroastrians living in colonies in Yazd, Kerman, and Tehran. It is thought that many Christians and Zoroastrians have left the country in recent years. The Jewish population numbered about 85,000 in 1974 but was the object of increasing persecution in the years leading up to the Islamic revolution. By 1980, so many Jews had left Iran that the Jewish population numbered only 32,000; persecution and emigration may have reduced that figure still further.

Daily Life

Iranians celebrate many holidays, both religious and political. The biggest is No-Ruz, or the Persian New Year, which falls in March. One No-Ruz tradition, that of leaping over a bonfire to burn away the bad luck of the past year, is probably a remnant of Zoroastrian fire worship. Political holidays include Oil Nationalization Day (March 20), Islamic Republic Day (April 1), and Revolution Day (June 5). Religious holidays are governed by the lunar calendar, so their dates change from year to year. The major ones honor the birthdays of famous *imams*, as the great spiritual leaders of early Islam were called. The birthday of the 12th imam, a special hero to Iranian Shiites, is celebrated with lights and candles all over the the country. Like Muslims everywhere, Iranians also honor the birthday of the prophet Muhammad and observe a month-long religious holiday called Ramadan, during which it is forbidden to eat, drink, or conduct business until the sun has set each day.

The most popular sports in Iran are wrestling, horse racing, and weight lifting; only men engage in them. Other sports such as soccer and volleyball achieved some popularity under the shah but have not been encouraged by the revolutionary government. Western movies and music are banned as well, although many Iranians, like young people the world over, are fascinated by American culture. Despite hard-line disapproval, imported Barbie dolls are one of the most popular toys.

Similarly, Western fashions and hairstyles were widely imitated in Iran before 1979 but are frowned upon today. In the cities, girls who wear makeup or boys who wear blue jeans or long hair are likely to wind up in jail or have stones flung at them by righteous mullahs. Girls and women are expected to wear the chador outside their homes; the preferred color is black. Men are expected to dress in simple, sober clothing, although it has been impossible to stem the tide of T-shirts and mass-produced Western-style clothing flowing from factories in Europe and Asia. Some of the mountain people still wear sheepskin robes, embroidered vests and skirts, and other forms of traditional dress.

Iran's position at the crossroads of trade and migration routes has contributed to its cuisine. Iranian cooks have borrowed the kabob (bits of meat skewered on a stick and grilled over charcoal) from Turkey; *dolmeh* (a paste of meat, rice, and peas wrapped in grape leaves) from Greece; curries (spicy stews served with rice, raisins, and relishes) from India; and many dishes using figs, dates, and lamb from Arabia. Iran's contributions to the world's tables include sherbet, melons, and kumquats. Fish is an important source of protein in the coastal regions, although more seafood is exported or converted into animal fodder than is consumed by the Iranians.

Although the shah's White Revolution encouraged urbanization, many Iranians continued to farm or to herd livestock, including this man, who brought his sheep to market in Tehran. This photograph shows how most Iranians, under the shah, wore Western-style clothing.

Chicken, often prepared with a sauce made from green plums, is a Kurdish specialty. Iranians love bread. Flat, round loaves are standard, but bakers often specialize in elaborate shapes, such as breads in the form of woven baskets. *Mast*, a dairy product much like yogurt, is popular throughout the country. Tea, which is grown in the Caspian region and on some of the mountain slopes, is the national beverage. The consumption of alcohol is strictly forbidden by Islamic law.

Contrasts between old and new, such as this chador-clad woman in front of a fried chicken restaurant, are common. The mullahs stress the traditional aspects of Iranian and Muslim culture but are unable to eradicate Western influences completely.

Villages and Cities

About 70 percent of Iran, chiefly the deserts and the more inhospitable mountain areas, is uninhabited. The western and northern parts of the country are the most populous, and the heaviest concentrations of people are in Tehran, along the Caspian coast, and in Azerbaijan. Although for many centuries Iran was an agricultural land, today more than half of the population dwells in urban areas. Iran's cities are growing so fast, as country people arrive looking for education and jobs, that housing and sanitation are inadequate in most communities. This trend is expected to continue, because many of Iran's young people are no longer interested in rural life and Iran's population is overwhelmingly young—about

45 percent of all people are under age 15. Another 31 percent are 15 to 29 years old; 11.5 percent are 30 to 44; 9 percent are 45 to 64 and only 3.5 percent are 65 or older.

The shah started many building programs designed to ease the housing shortage. As a result, high-rise apartments and other modern, Western-style dwellings appear throughout Iran, primarily in the larger cities. In several regions, however, traditional architectural styles still prevail. Along the Caspian coast, many people live in small, scattered villages of two-story wooden houses with shuttered windows, surrounded by sturdy henhouses and barns. In the mountain villages, which are situated on rocky slopes overlooking the pastures and terraced fields, houses are square and made of mud brick. They do not have windows, but holes in their flat or domed roofs let in air and light and emit smoke. Villages on the plains tend to be organized around a central square in front of a mosque. The high mud-brick walls of the houses have corner towers, and the flat straw roofs, supported by wooden rafters, provide places to dry fruit and store provisions.

Iranian cities follow the Islamic custom of having separate districts for businesses, government offices, and residences. Although the government has built broad thoroughfares and encircling highways to carry modern traffic, the heart of each city is still a labyrinth of narrow, twisting streets and squares, many of which are unnamed. Each city has a bazaar, or marketplace, where tradesmen and artisans carry on their business from small shops or rugs spread on the ground. Within the bazaar, merchants are grouped according to what they sell. For example, all the silversmiths are in one area, all the leatherworkers in another, and all the spice vendors in yet another. Some of the bazaars are quite ancient and very large. The bazaar of Tehran is a covered maze that contains more than 6,000 stalls and 9 miles (14.4 kilometers) of walkways. It is filled with the smoke of portable charcoal braziers (Iran's most common heating and cooking device) and the smell of flowers, foods, perfumes, and animals. Shiraz's bazaar is a complex of yellow brick passageways with high, gracefully arched ceilings that echo to the din of metalworkers' hammers and vendors' shrill cries.

Tehran, the capital and largest city, is growing at a tremendous rate. It now has almost 9 million inhabitants. Because it was a fairly insignificant city until it became the Qajar capital in the late 18th century, it does not have as many impressive old buildings as some other Iranian cities. It does, however, have the Gulistan Palace, which was the shah's residence, and a number of mosques and government buildings erected by the Pahlavis.

Mashhad, Iran's second-largest city, was an important post on the ancient Silk Road. It is holy to Shiite Muslims as the burial place of Ali Reza, the eighth Shiite imam, and many pilgrims visit the shrine at his grave. A mosque that was built by Queen Gauhar Shah Khatum, a daughter-in-law of Tamerlane, is generally regarded as

The traditions of Persian excellence in literature and calligraphy are combined in this illustrated page from an epic by Nēzamī. Calligraphy and miniature painting reached the height of accomplishment under the Safavids, who ruled when this work was produced in the 16th century.

the most beautiful building in Mashhad and one of the loveliest in Iran. It is ornamented with magnificent mosaics in brilliantly colored tile. Nadir Shah, who made Mashhad his capital, built an imposing shrine and tomb for himself there.

Esfahan, Iran's third-largest city, has long been famous as the home of peerless craftsmen and artists. Shah 'Abbas made it his capital and built a number of grand structures, including the Maidan-i-Shah, an enormous open square that served as a parade ground for his troops in the center of the city, and the Masjid-i-Shah, or royal mosque. Esfahan is famous for the vivid blues of the tiles with which the Masjid-i-Shah and other buildings are roofed; from a distance, the city looks like a stream of turquoises tumbling out of the brown hills amid which it is set. The city of Shiraz, to the south, has a mosque that dates from the 9th century. Although it has fallen into ruins, it is being restored. Shiraz is also the burial place of two of Iran's greatest poets, Sa'di and Hafez. The silverwork and rugs of Shiraz's artisans are renowned throughout Iran.

The Iranian Arts

Since the time of the Sassanid goldsmiths, the Iranians have been regarded as an artistic people. Metalwork in gold and silver continues to be practiced today, and ornamental cups, bracelets, and earrings are made to ancient designs. Iran also possesses a fabulous treasure in gems, the former crown jewels, now the property of the state. This collection, which includes the famous Peacock Throne, studded with more than 20,000 precious stones, is probably the single most valuable set of gems in the world. It is kept heavily guarded and serves as the backing for Iran's currency. Many of the crowns and other priceless pieces in the collection were made by Iranian artists, but others were looted from the Moghul palaces and elsewhere by Persian conquerors.

Islam brought a new artistic tradition to Iran. Calligraphy (skilled lettering) of the Koran and other religious works became an

art form. It was accompanied by painting. Although strict interpretation of Islamic doctrines forbids the painting of human or
animal images, this rule was never really observed in Iran, and the
Koran was sometimes illustrated. After the traditions of Chinese
landscape painting were imported by the Mongol conquerors,
bookmaking became an Iranian specialty. Hundreds of manu-
scripts were produced with fine swirling calligraphy and bright
pictures of gardens, warriors, magicians, and princesses. Minia-
tures, as these small illustrations are called, were prized for their
remarkably fine detail and jewellike colors. The art of miniature
painting reached its highest expression in the second part of the 15th
century, when an artist named Bihzad created masterpieces. The
few examples of his skill that survive are among the world's costliest and most prized artworks.

Iran has a long and rich literary tradition, especially in poetry.
Some fragments of *qasidas* (formal odes) and *ghazals* (love poems)
from the 9th century have come to light, but Iranians generally agree
that the first of their great poets was Firdawsī (940–1020), whose
120,000-line *Shahnameh* (Book of Kings) is Iran's national epic. It
recounts the half-fictional, half-historical adventures of four dynasties of ancient Persian rulers. These tales are still popular and
universally known in Iran today. The *Shahnameh* was copied and
illustrated by many of medieval Iran's finest painters and callig-
raphers.

Omar Khayyám, who died in the first third of the 12th century,
wrote about 1,000 quatrains (four-line poems) called *ruba'is*. He
became famous around the world centuries later when the British
writer Edward FitzGerald (1809–83) translated 101 of the ruba'is
and published them in a volume called *The Rubáiyát of Omar
Khayyám*. Although not all of FitzGerald's translations were strictly
faithful to the original Persian, they became immensely well known.
One of the most famous of them runs:

The Moving Finger writes, and, having writ
Moves on; nor all your Piety nor Wit
Shall lure it back to cancel half a Line,
Nor all your Tears wash out a Word of it.

Another famous poet, Nēzamī, who died around 1203, composed five epics, including one about the exploits of Alexander the Great in Iran.

Later writers included Saʿdī (1184–1291), among whose works were *Golestan* (The Rose Garden) and *Bustan* (The Orchard). Hāfez, who lived about 100 years later, wrote some 700 love poems. Poets of the 20th century include Iraj Mirza, Parveen Ettasami, and Behar. Modern Iranian writers, influenced by Western culture, have also written novels. Among the best-known novelists are Zain-al Abidin, who wrote tales of social criticism and protest in the early years of the century, and Sadiq Heyat, author of *Buf i kur* (The Blind Owl), the first Iranian book to draw upon the speech and customs of the peasants.

Iran's most famous art form is probably the handwoven carpet with a pattern of geometric shapes or Arabic lettering. Indeed, the term *Persian carpet* is often used to refer to such rugs, whether they are made in Iran, India, Pakistan, or Turkey. Each region of the country, or each city, has its own traditional carpet designs and colors; an expert can easily tell a Shiraz carpet from a Tabriz one, for example. Collectors around the world cherish Iranian rugs, and some of the most prized are centuries old. Although many rugs are now machine-made for export, old-fashioned hand looms are still to be found in most cities and villages.

About 33 percent of all Iranians work in agriculture. Among Iran's major crops are wheat, barley, rice, sugar beets, raisins, dates, and tea. These women are collecting tea leaves on a plantation.

7

Resources and Economy

Iran's most valuable resources are petroleum and natural gas. In fact, petroleum has been the lifeblood of Iran's economy for most of the 20th century. Experts estimate that Iran has about 9 percent of the world's known oil reserves. In that respect, Iran is among the top five countries in the world. Along with the nations of the former Soviet Union, Iran also has the largest reserves of natural gas. Most petroleum and gas development has occurred in the southwest and in the offshore waters of the Persian Gulf, but oil has also been drilled from the region near Qom and in the Dasht-e-Lut. In the early 1980s, Iran was the second most productive nation in OPEC (the Organization of Petroleum Exporting Countries, a cartel, or group that decides together the amount and the price of oil its members sell). But the long war with Iraq caused a drop in the productivity of Iran's oil fields and refineries. Furthermore, despite OPEC's efforts, after 1985 the international price of oil fell, so Iran's income from the sale of its petroleum decreased still more. Although the government has encouraged the development of new industries and other resources so that Iran will no longer be dependent upon a single resource, it is likely that oil will remain Iran's principal commodity for years to come.

Camels, an age-old method of transportation across Iranian deserts, graze in front of a modern cement plant. Cement production is one of Iran's more important industries.

But most of the oil and gas is exported or used to operate manufacturing plants. Fuel for cooking and heating within the country is sometimes scarce. Charcoal braziers (grill-like heaters), wood fires, and coal stoves are widely used. Since the 1960s, however, a number of hydroelectric power plants have been built to tap the energy of Iran's fast-moving mountain streams.

The country possesses solid mineral resources, too. Its copper reserves, mostly in Azerbaijan and Kermanshah, are estimated to total 1 billion tons. Lead, zinc, coal, sulfur, and chromite occur in significant amounts. Iron ore, manganese, and salt have also been mined. A mining law that was passed in 1957 divides Iran's mineral resources into three categories. The first is building materials such as stone and marble. These may be used by private landowners. The second includes metal ores, precious stones, salts, and coal; landowners must obtain licenses from the government and pay high taxes to exploit these resources. The third category—oil, gas, and radioactive substances—belongs entirely to the government wherever it is found.

About 21 percent of Iran's work force is employed in industry. In addition to the petroleum and mining industries, this group works in steel mills and in plants that produce automobiles, television sets, textiles, cement, fertilizer, and processed foods. The government has encouraged the growth of the paper, shoemaking, pharmaceutical, aircraft, and shipbuilding industries in the hope of making Iran less dependent upon goods imported from other countries.

Iran's major trade partners are Japan, Germany, the Netherlands, the United Kingdom, Italy, Spain, Turkey, and France. It exports mainly petroleum, but also such items as carpets, fruits, nuts, and hides; it imports machinery, food, pharmaceuticals, and military supplies.

Agriculture employs 33 percent of the work force. Nearly 10 percent of the country's area consists of cultivated farmland; another 12 percent is used as pasture. The chief agricultural products are wheat, barley, rice, sugar beets, cotton (used in the textile industry), raisins, dates, tea, and tobacco. Almonds and pistachio nuts are grown for export. Farming villages as well as the nomadic tribes keep livestock. Sheep and goats are most numerous, but Iran also has large numbers of cattle, water buffaloes, and horses, as well as about 24,000 camels. The most important animal products are wool for carpets and textiles, milk, butter, hides for the leather industry, and animal fat for cooking. Most cattle are kept for milk or for pulling loads and are not fattened for beef.

Forestry and fishing are also segments of Iran's agricultural economy. The government controls the timber industry, which is concentrated in the Elburz Mountains. About 12 million hectares (29.7 million acres) of timberland are available for harvesting. Most of this wood is used for building. Fishing is concentrated on the Caspian coast, because the Iran-Iraq War and oil development have harmed the Persian Gulf fisheries. The total catch of Iran's fisheries is probably around 20,000 tons, mostly sturgeon and salmon. Iranian caviar, nearly all of which is exported for the large sums in foreign currency that it brings into the country, amounts to perhaps half of the world's total supply.

Iran's unit of currency is the rial, which is divided into 100 dinars. In the late 1990s, the average income of an Iranian worker was equal to about $1,500 in U.S. money—a sum roughly one-third that of two decades earlier. As this sharp decline in income suggests, Iran has experienced a spate of economic hardships since the 1970s. Shortages of fuel and food have become common in many parts of the country. Plants and businesses have been forced to close for lack of money to operate or customers to purchase goods. Experts who have studied recent events in Iran speculate that unemployment is rising and that as much as 30 percent of the work force may be out of work.

Transportation and Communications

Historically, travel between Iran's widely distributed centers of population was made difficult by its mountains and deserts. One of the priorities of the Pahlavi regime was to improve and modernize the country's transportation system.

Today Iran has about 87,120 miles (140,200 kilometers) of roads; about one-third of them are paved. By 1995, the country had about 1.6 million passenger cars and 600,000 trucks and buses. These vehicles are the primary means of transport for both people and materials. The country has only 3,165 miles (5,093 kilometers) of railroads. The main line runs between the Caspian Sea and the Persian Gulf, passing through Tehran on its way. A railway connection at the Turkish border links Iran to Europe, and a line that reaches Mashhad may someday be part of a planned railway across Asia.

Ferries carry passengers and cargo across Lake Urmia and between ports on the Caspian coast. Ships also cross the Caspian to trade with nations of the former Soviet Union. Ports at Khorramshahr, Kharg Island, and Abadan on the Persian Gulf were damaged in the Iran-Iraq War. The state-owned national airline is called Iran Air. It operates from international airports at Tehran, Esfahān, Shiraz, and Bandar ʿAbbās. Local airlines serve smaller airports in many other cities and towns, and some European, African, and Asian airlines use the larger airports.

Oil is the most important natural resource in Iran. The nation's economy depends on oil and gas exports, although most of the oil is refined elsewhere.

Roads are the primary means of transport for people and goods in Iran, although only one-third of the roadways are paved. These Iranians did not have access to one of the nation's 1.6 million cars but managed to find another means of motorized transport.

Most communication within Iran is by radio. The country has a radio set for every 5 people, compared to 1 television set for every 26 people and 1 telephone for every 25 people. The telegraph, telephone, radio, and television services are owned by the state. All broadcasts are censored by government and religious officials. It is against the law to broadcast anything that may appear critical of Islam.

In 1979, the revolutionary government issued a new press law that required all newspapers to be licensed by the state and said that editors and reporters could be fined or sent to jail if a newspaper criticized Islam, the government, or a religious leader. Many of the country's 39 papers were shut down. (The previous regime, under the shah, had also exercised strict censorship and repression of the press.) In 1985, Khomeini announced that censorship and other restrictions on freedom of the press would be loosened, but it does not appear that much change has taken place. Iran now has 36 daily papers, most of which are published in Tehran. Their total circulation is about 2 million. The parameters of what is permitted tend to shift quickly, in response to pressures within the ruling movement. What can be said, written, or filmed today could be cause for financial ruin or arrest tomorrow.

The staunchly conservative Iranian speaker of parliament Ali Akbar Nateq-Nouri, pictured here, was soundly defeated by Mohammad Khatami in the 1997 presidential election. Although Nateq-Nouri had the support of Iran's conservative Islamic establishment, Khatami won a surprise victory due to the huge turnout of Iranian voters who favored the more moderate course that Khatami seemed to represent.

8

A Country at a Crossroads

Iran's position between Africa, Asia, Arabia, Europe, and India has deeply affected this mountainous land throughout history. Migrant peoples in search of new homelands, traders in search of profits, religious teachers in search of believers, and generals in search of new conquests have swept across Iran, each leaving something to shape Iranian culture.

Yet Iran's rugged geography and the fierce independence of its many peoples also kept the country somewhat isolated. Today its isolation is self-imposed. The shah propelled a farming people with a centuries-old way of life into the 20th century in less than a generation. But with one foot in the modern world, Iran revolted against Westernization and modernization. The Islamic revolution, while it has attempted to keep Iran moving forward economically, turned the cultural calendar back by decades. Iran became an experiment in Islamic fundamentalism, a country whose leaders believed that the best way to go forward into the future was to return to the traditions of the past, when all life was organized around unquestioned religious principles.

The outcome of this experiment is in doubt. Internally, Iran's troubled economy and repressive political system have sparked

On June 7, 1989, three days after the death of the ayatollah, mourners thronged a memorial service for the religious leader, holding his portrait aloft and shouting their grief. Although many observers feared that political chaos would follow Khomeini's death, Iran has weathered the storm and is pursuing a more moderate course than when it was under his leadership.

criticism of the revolutionary government and dissatisfaction with some of its policies. Ethnic minorities such as the Kurds and the Qashqais have openly rebelled. Externally, Iran has failed to achieve a position of leadership among the other Islamic nations of the world, perhaps because of its Shiite faith––or perhaps because other powerful Muslim countries, such as Saudi Arabia, Indonesia, and Egypt, are aware of the economic and political hazards of becoming an enemy of the United States.

In the early 1990s, Iranian leaders agreed to negotiate with the United States to free Western hostages held in Lebanon, in exchange for the release of Iranian funds held by U.S. banks. This cooperation led to the last of the Western hostages going home in 1992. But the Persian Gulf War, in which Iran condemned both Iraq's invasion of Kuwait and the West's intervention to rescue Kuwait, disturbed relations anew. In 1995, the United States banned all trade with Iran, condemning the country as a "rogue regime." A 1997 German court verdict linked the murders of four dissidents to Iran's top leadership, seriously damaging already limited diplomatic ties between Iran and the European Union. For its part, Iran has turned to the East to forge stronger trade ties. China and Russia, which are already the largest arms suppliers to Iran, are the most likely partners in the near future.

When Mohammad Khatami succeeded Rafsanjani as president in 1997, Iran entered a new phase in economic development and international relations. Because of Khatami's emphasis on international trade, civil liberties, and political pluralism, many viewed his election as a major setback for the hard-line establishment. But whatever moderation Khatami may exercise in domestic affairs, experts assume he will have little power to alter foreign policy. It remains to be seen how the balance of moderates and conservatives will shift in the future.

GLOSSARY

anjuman   Provincial council.

ayatollah   Literally, "sign of God." Used as a title of respect for a religious leader.

Majlis   The Iranian legislature; the National Consultative Assembly.

mullah   A learned man or community religious leader.

ostan   An administrative province.

qisas   Part of the penal code that allows retributive punishments such as amputations or executions.

shah   King; the emperor or king was sometimes called the shahanshah, or king of kings.

Shia   A form of Islam practiced by the Shiite Muslims, who make up about 10 percent of the worldwide Islamic community. Shia split from majority, or Sunni, Islam centuries ago in a dispute over the succession to the position of heir to Muhammad's religious and political leadership.

Shiite   See Shia.

ulema   Council of high-ranking religious authorities.

valiy-e faqih   The position of supreme religious leader created by Iran's 1979 constitution and occupied by Khomeini from 1979 to 1989.

INDEX

A

Abbās I, Shah, 49, 93
Abidin, Zain-al, 95
Achaemenids, 39–42
Afghanistan, 21, 22, 39, 42
Afshar dynasty, 51–52
Agriculture, 96, 99
Ahura Mazda, 41, 47
Alexander the Great, 15, 42–43
Animal life, 33–35
Arabs, 16, 22, 46, 47, 85
Armed forces, 74
Armenia, 21, 85
Armenians, 85, 87
Aryan migration, 38–39, 81
Azari, 81
Azerbaijan / Azerbaijani, 21, 22, 28, 31, 32, 48, 54, 81–82, 85, 87, 90

B

Bactrians, 39
Bakhtiari, 23, 82
Baluchis/Baluchistan, 28, 84
Bandar ʿAbbās, 24, 101
Bazaars, 91–92
Bihzad, 94
Birds, 34–35

C

Calligraphy, 93–94
Carpets, 16, 49, 95
Caspian Sea, 21, 23, 25, 27, 29, 32, 101
Censorship, 103
Central Elburz National Park, 33
Chador, 53, 54, 66, 84, 88
Christians, 46, 66, 87
Cities, 24, 25, 90–93
Civil rights violations, 78–79
Climate, 28–32
Communications, 102–3
Constitution, 71–72
Ctesiphon, 43–44, 46
Currency, 100
Cyrus II (Cyrus the Great), 39, 40

D

Damavand Mountain, 26
Dasht-e-Kavir, 24
Dasht-e-Lut, 24
Deserts, 24
Dress, 88
Druze, 87

E

Earthquakes, 27

110

Ecbatana (Hamadan), 39
Education, 75–77
Elamite kingdom, 15, 27, 38
Elburz Mountains, 24–27, 29, 100
Esfahan, 24, 29, 38, 49, 81, 85, 93, 101
Ettasami, Parveen, 95
Euphrates River, 27, 38

F

Fars (Parsa), 39, 44
Farsi, 81
Fashion, 88
Firdawsi, 94
Fishing industry, 100
FitzGerald, Edward, 94
Food, 88–89
Forests, 32–33

G

Genghis Khan, 15, 47, 48
Ghaznavids, 47
Ghotbzadeh, Sadegh, 66
Greece, ancient, 15, 27, 39–40, 42
Gulf of Oman, 21, 28

H

Hafez, 95
Hasan-e Sabbah, 46
Hashishi, 46
Health care, 77
Heyat, Sadiq, 95
Holidays, 87–88
Hostage crisis, 17, 66
Housing, 91
Hydroelectric power, 29, 98
Hyrcania, 27, 28, 32, 42, 43

I

Il-Khanid dynasty, 48

Industry, 99–100
See also Oil industry/resources
Iran
daily life in, 87–89
economy of, 99–100
emergence of civilization in, 15, 37–38
ethnic groups of, 23, 81–85
geographic features of, 23–28
government system of, 71–73
Iran-Iraq War, 17–19, 66–68, 85
judiciary system of, 73–74
location of, 21–23
post-Khomeini, 19, 68–69, 107
prehistory of, 37
religious groups in, 46, 66, 87
Iraq, 21, 22, 82
Irrigation, 29–30
Islam(ic), 22–23, 49, 86–87, 92
conquest, 15–16, 22, 45–47
Revolution, 16–17, 65–66, 105–6
Islamic Revolutionary party (IRP), 65, 66, 75, 79
Isma'iliyah, 46

J

Jews, 46, 66, 87

K

Karun River, 29
Khamenei, Ali, 19, 72
Khamsheh, 85
Khan, Agha (Mohammad), 52
Khan, Hulagu, 47–48
Khatami, Mohammad, 19, 69, 107
Khayyám, Omar, 94–95
Khomeini, Ayatollah Ruhollah, 16–17, 19, 56, 65, 66–67, 68, 71, 72, 77, 79, 103

Khorasan, 28, 32, 52, 82, 85
Khorasan Mountains, 24, 27, 28, 32
Khuzistan, 27, 31–32, 37, 66, 67, 85
Kurds, 82

L
Lake Urmia, 28, 101
Language, 81, 82, 84, 85
Luristan, 23, 28, 82
Luri (Lurs), 82

M
Majlis, 53, 55, 56, 72
Makran (coastal plain), 28
Mamasani, 85
Mashhad, 27, 47, 52, 92–93
Medes, 39
Metalwork, 16, 44, 93
Mineral resources, 98–99
Miniature painting, 94
Mirza, Iraj, 95
Mithradates I and II, 43
Mithraism, 43
Mongol invasion, 47–48
Mosaddeq, Mohammad, 55–56
Mountains, 24–27
Muhammad, 45–46, 86, 88

N
Nadir Shah, 51–52, 93
Natural gas, 97
Newspapers, 103
Nēzamī, 95
Nomadic tribes, 31, 82–84, 85

O
Oil industry/resources, 53, 54, 55–56, 85, 97–98, 99

OPEC (Organization of Petroleum ExportingCountries), 97

P
Pahlavi, Mohammad Reza Shah, 16, 17, 54–56, 65, 77, 78, 91, 105
Pahlavi, Reza Shah, 52, 53–54
Pakistan, 21, 22, 28, 39
Parsa (Fars), 39
Parsis, 47
Parthians, 39, 43–44
Peacock Throne, 50, 93
Persepolis, 40, 41, 42
Persia, 15, 39, 54. *See also* Iran
Persian Gulf, 21, 27, 29, 101
Persians, ethnic, 81
Plant life, 32–33
Poetry, 94–95
Political parties, 75

Q
Qajars, 52–53
Qanat system, 29–30
Qashqai, 85
Qisas system, 74
Qom, 24

R
Rafsanjani, Akbar Hashemi, 19, 69, 71, 73, 107
Railroads, 53, 100–101
Rivers, 29
Roads, 100
Roman Empire, 16, 41, 43, 44–45
Rubáiyát of Omar Khayyám, The, 94–95
Rushdie, Salman, 19

S
Sa ʿdī, 95

Safavids, 48–49, 51
Safid River, 29
Sassanids, 44–45
Satanic Verses, The (Rushdie), 19
Saudi Arabia, 16, 22, 66
SAVAK, 56, 65, 78
Seleucids, 43
Seljuk Turks, 47
Shapur I, 44
Shatt-al-Arab, 27, 29, 66
Shiite Islam, 23, 46, 49, 86, 92
Shiraz, 25, 81, 92, 93, 101
Silk Road, 16, 24, 43, 92
Social-welfare benefits, 77–78
Soviet Union, 18, 21, 54–55, 101
Sports, 88
Strait of Hormuz, 21, 24
Sufism, 87
Sultan Husein, Shah, 51
Sunni Islam, 46, 86
Susa, 38, 40, 41, 43

T
Tabriz, 25, 48
Tehran, 17, 24, 26, 29, 81, 85,
    90, 91–92, 101
Tigris River, 27, 38
Timber industry, 99–100
Timurid dynasty, 48, 49
Trade, 16, 27, 99, 101

Transportation, 100–101
Turkey, 15, 21, 22, 39, 47, 82
Turkmenistan, 21, 22, 85
Turkomans, 85

U
United States, 17, 18, 54–55, 56,
    66, 107

V
Valiy-e faqih, 72
Volcanoes, 26–27

W
Women, 53, 54, 56, 66, 75, 77,
    83–84, 88
World War I, 53
World War II, 54

X
Xerxes I, 39–40

Y
Yazd/Yazd region, 28, 29

Z
Zagros Mountains, 24–25, 27, 28,
    29, 32, 82
Zands, 52
Zayandeh River, 29